9.22

THE CATHOLIC

THE CATHOLIC FAITH

RODERICK STRANGE

Oxford   New York

OXFORD UNIVERSITY PRESS

Oxford University Press, Walton Street, Oxford OX2 6DP

Oxford New York Toronto
Delhi Bombay Calcutta Madras Karachi
Petaling Jaya Singapore Hong Kong Tokyo
Nairobi Dar es Salaam Cape Town
Melbourne Auckland

and associated companies in
Beirut Berlin Ibadan Nicosia

Oxford is a trademark of Oxford University Press

First published 1986 as an Oxford University Press paperback
and simultaneously in a hardback edition

Paperback reprinted (twice) 1986

British Library Cataloguing in Publication Data

Strange, Roderick
The Catholic faith.
1. Catholic Church—Doctrines
I. Title
230'.2   BX1751.2
ISBN 0-19-826685-5
ISBN 0-19-283051-1 (Pbk)

Library of Congress Cataloging in Publication Data

Strange, Roderick
The Catholic Faith.
Bibliography: p.
Includes index.
1. Catholic Church—Doctrines—Addresses, essays,
lectures.   I. Title
BX1755.S78   1986   230'.2   85-29689
ISBN 0-19-826685-5
ISBN 0-19-283051-1 (pbk.)

Nihil obstat:

DAVID MCLOUGHIN, STL

Imprimatur:

✠ MAURICE COUVE DE MURVILLE
Archbishop of Birmingham

Printed in Great Britain by
Richard Clay (The Chaucer Press)
Bungay, Suffolk

Preface

As a university chaplain I have often needed to hand to an enquiring student—sometimes a Catholic, sometimes not—a book that gave an account of the Catholic faith which was readable and not too long, intelligent, but not too technical. I could never find one. What follows is my own attempt to supply that need. Its origins are explained more fully in the first chapter, but it took shape most recently from lectures given to Oxford undergraduates. Nevertheless, more exactly it comes from my experience of life as a Catholic priest: from celebrating sacraments and praying, from preaching and teaching, from loving and being loved, from trying to share in the lives of friends and those who have come to me in need, and also, of course, from my failures and defeats. I have been personal if necessary, but only in order to present the Church's teaching more clearly. That at least has been my intention. I hope that the result will be a help not only to readers, but also to priests and religious who may have been looking for such a book, as I have been.

It might be helpful to state the approach of this book very briefly. It is based on three main themes. The first is the belief in Jesus of Nazareth as truly God and truly a man. In other words, in him the divine and the human are related most perfectly and so the relationship of the divine and the human elsewhere, in the Church, in the scriptures, in the sacraments, and so on is to be understood in accordance with that perception of him as truly God and truly a man. Secondly, when that relationship of the divine and the human is being realized amongst us most effectively, we often find, because of sin, that we are placed in a crisis which must be overcome: this experience draws us into a sharing in the death and resurrection of Jesus. And thirdly, we have to reflect upon our share in the relationship of the human with the divine and on our experience of death and resurrection in order to gain the self-knowledge which is essential for us as Christian men and women and as a Christian community. You will find that these three themes emerge and intermingle constantly in what follows.

The section on the immaculate conception appeared as an article,

79494

'Immaculate Mary', in the Catholic newspaper *The Universe* on 7 December 1984. Part of the chapter on the Trinity has appeared in *The Clergy Review* with the title, ' "Relations are Real": A Sermon on the Trinity' (no. LXVIII, May 1983, pp. 179–80). I am grateful to both editors for permission to reproduce the material here. Scriptural references have been taken from the Revised Standard Version, and quotations from the documents of Vatican II come generally from the edition by Walter Abbott (London, 1966).

My debts of gratitude are beyond counting. However, I mention gladly Fr. Reginald Riley, SM, and Fr. Edward Yarnold, SJ, who gave me valuable advice on the typescript. Fr. Timothy Radcliffe, OP, then went through it and offered me a wide range of suggestions which I have tried to use wherever possible. He deserves my special thanks. Jill and Frank Harkins, my elder sister and her husband, let me use their Lake District cottage to begin writing the book in September 1984 and Frank and Mary Ashby gave me vital respite during the spring of 1985 to complete it at Folly Bridge. Their kindness has been indispensable for me. Christine Budgen has typed my handwritten script without a murmur of complaint. I never cease to be amazed at her patience and generosity. And at Oxford University Press, Anne Ashby and Hilary Feldman have been my allies and given me their unfailing help. I am most grateful to them. Finally, I owe an immense debt to my family and friends for their love and support and to all those who, by asking for help, have taught me so much. I thank them all.

The Old Palace   RODERICK STRANGE
Oxford
Feast of St Teresa of Avila, 1985

Contents

I

Faith seeking understanding

(i)

WHEN I came to the University Catholic Chaplaincy in Oxford in 1977, one of my tasks for the first term was to lead a series of discussions on faith. The sessions followed a well-tried pattern: I would speak for about half-an-hour; there would then be questions for forty-five minutes or so; and finally we would break off for coffee and to talk more informally. On the first occasion, over coffee, a girl came up to me and thanked me warmly for speaking as I had done. She was most generous in her comments. And amidst the praise she was critical of her earlier experience. 'We were never given anything like that at school,' she said. There was a pause and honesty struck as she added, 'Mark you, if we had been, we wouldn't have listened.' Perhaps the seed of this book was sown in that conversation.

The character and style of religious education have changed out of recognition over the past twenty-five years. The emphasis on the catechism and learning by heart has largely passed away. We need not waste much time in regret. I heard recently of a teacher who tested his sixth-formers in their catechism rather more than twenty years ago. The change was then just beginning, and fearful, protesting voices were being heard, challenging the newer methods. He discovered that what these boys remembered from the catechism which they had been learning from their earliest years was very slight indeed; and what they understood of what they remembered was minimal. Of course, there must be those who learnt a great deal through the old method, but they are perhaps people with the disposition to have learnt much from any method. No, we need not regret the passing of the Penny Catechism.

The newer method itself, however, is no panacea. Its aim is to make possible that personal relationship with God which bears fruit in commitment and love. Obviously there is no recipe, no formula which can be applied, that can guarantee that result. It is a question of trying to create a setting in which the discovery of God in love and knowledge and delight can take place. There are bound to be failures. That is nothing new. There were many failures before, but they were disguised more easily by the knack of rattling off the answer to a question from the catechism. Nowadays, when God has not been found, his absence is more starkly evident. And there is another consequence of this newer approach; opinion will vary as to whether it is a drawback or not.

When it works well, this way of helping people to come close to God stirs in them even from an early age a wonderful depth of devotion and prayer, as well as sensitivity to the needs of others. But it tends to lack intellectual content. Dealing with university students, chaplains often comment that they find their intentions and religious dispositions admirable, but their knowledge rather slight. And so we are back to the purpose of this book and my conversation with that undergraduate in October 1977.

One of the duties of a university chaplain is to be available to discuss the queries that students may have about their religion. Sometimes the questions are raised by people who are not themselves Catholics. Perhaps they wish to be; maybe they do not. Much of the material in these pages has taken shape from discussions of that kind. I hope the book will be useful for people curious about Catholicism and who are beginning to look for information. Many other enquirers, of course, will be Catholics already. A particularly encouraging feature of my work as a university chaplain has been meeting large numbers of young people aware of their lack of knowledge and eager to overcome it. On many occasions people have come up to me to say they were having a beer with someone the previous evening and had been asked about papal infallibility or the immaculate conception or some other religious matter. So often their comment has been, 'I know what I believe, but I didn't know what to say.' What follows is a response to that feeling. And, of course, the predicament is not the exclusive preserve of undergraduates.

There are, for example, those of my own generation. Leaving school in the late fifties or early sixties and bound deeply to their religious faith which they had received as something unchanging, solid, and secure, they were establishing themselves in their careers when the Second Vatican Council struck the Church. Some appreciated what was happening, some were appalled, and many were bewildered. They remained loyal in their believing, but, being confused, ducked their heads, and decided not to think about such matters any more. It would be good to think that the appreciative might find here strength, the appalled reassurance, and the bewildered, particularly the bewildered, something which sheds light on their perplexity, so that they can enjoy the more what they already believe.

Then there are those who are older, the parents, as it may be, of today's students. Any chaplain has dealings with them and finds amongst them those who at little prompting display anger and distress because they feel that they have been betrayed. Their children, they judge, seem to lack the knowledge of their religion which they wish for them; and yet on investigation the knowledge that they are advocating proves to be alarmingly superficial. It rests largely on an understanding of the Pope and his authority which tends to be as extreme as it is over-simple. Their attitude is marked by fearfulness. I would be happy to relieve them of such an unwelcome burden.

There is one other group I would like to mention. They are older still, more mellow, but perhaps saddened. They received their understanding of their faith from people like my most famous predecessor, Ronald Knox. It has stood them in good stead, but something is missing for some of them. A friend of mine once described his father to me as a fine example of that generation of Catholics; 'but', he noted, 'he lacks one thing: he was never prepared for the possibility of change'. It would be a pleasure to supply that lack.

These are the people I have in mind as I write. They range from eighteen to eighty, and perhaps beyond. So let me try to describe in a little more detail how I plan to proceed.

(ii)

Before my appointment to the Oxford Chaplaincy, I spent three years at English Martyrs' parish in Wallasey on Merseyside and was also chaplain to St Mary's College, the Catholic comprehensive school in the town. In the spring of 1977, Peter Hughes the headmaster and I arranged a series of evenings for parents, to acquaint them with the kind of religious instruction their children were receiving. During one session I was asked the key question. A father stood up: 'You have been telling us things about the Church and morality and the sacraments. But when we were at school twenty years ago we were being told different things. How do we know that in twenty years' time another priest isn't going to be here, giving quite different answers again to these questions?' That seemed to me to sum up succinctly the difficulties and anxieties which many Catholics have felt at this time. They conclude from the changes that there are no reliable answers available. Of my reply to this charge I can recall only that on the spot it was vague and unconvincing. But I looked at the problem more carefully afterwards and came to realize what I should have said.

The flaw in the question lay in seeing the replies of 1957, 1977, and 1997 as simply comparable alternatives, as though, for example, what was a mortal sin in the past might be classified as venial in the present and possibly celebrated as a virtue in the future. But I had not been giving 'different answers', as a person might supply an alternative route to a friend going out for a drive from Wallasey to Wigan. It was rather a matter of sketching out the landscape afresh; an improved map might change the entire approach to the journey. So in that instance it was not a question of re-classifying particular moral acts, but of trying to reach a deeper and truer view of Christian morality. This book is an attempt to sketch that landscape; it will try to give an account of Christ and the Church and to indicate a response through the sacraments and in the Christian life of virtue.

I know that I have not given an exhaustive, systematic account of Catholicism. Much is omitted which some will no doubt judge essential to my purpose. Shortly before I began writing in earnest, I had lunch with a friend, Colin Wells, who had just written a book with similar intentions on *The Roman Empire* (London, 1984). We agreed completely on the difficulties of composition which are involved. Colin expressed the matter succinctly in his Preface: 'the professional scholar, writing in a field which has seen so much scholarly activity, feels his colleagues looking over his shoulder, and is obsessed with the need to justify himself, to reassure them that he realizes when he is over-simplifying, or when he has passed over in silence a topic generally deemed to be fascinatingly controversial, which generally means hopelessly obscure'. I am a pastoral priest, not a professional scholar. There will be points I have missed, because the demands on my time make my visits to libraries rare. But I hope that what I have done is not thereby undermined.

I have wanted to supply an account of Catholicism which will be of real use to people. I once asked a parishioner to explain to me in words of one syllable how a car engine works. He lost me in half a sentence and I was left wondering how often I had lost a congregation after the first half-sentence of a homily. I have tried to meet people at their point of difficulty. As I have been writing, the phrase that has kept coming back to me has been, 'Look at it this way.' My intentions have been orthodox. Idiosyncrasy would be misplaced here. And if I have at times said less than some people would wish, my reason has been consistent: to give an account of Catholicism which can be understood and which will not be obscured by preconceived difficulties. No one who reads on will interpret that remark rationalistically; hard sayings have not been bypassed; but I have tried to overcome unnecessary obstacles. I have wanted to offer people a starting-point for understanding their faith.

(iii)

I propose to adopt two main guidelines. First, in an attempt to know and love God, we should be confident of success; and secondly, the means of achieving such knowledge and love will generally (we cannot rule out exceptions) be familiar to us. Let me say a word about each.

1. As Christians we believe in a God whom we perceive as a personal being, all-knowing, and all-loving. Moreover, we recognize ourselves as his creatures, endowed with a capacity for knowledge and love. Now it may be possible that such a creator of such creatures could so design them as to be capable of knowing and loving but in fact utterly incapable of knowing and loving him, although he is a God of knowledge and love. But such a theoretical possibility seems to me to destroy itself on its own absurdity. Of course, if the Christian God is a delusion, there is nothing more to be said. But assuming that faith is properly founded, to assert that human beings are endowed with a knowledge and a love which is unavoidably in error and necessarily deceived is to deny any reality to knowledge and love at all. If we can know truth and love truly, which we are confident we can, and if we believe that our God who knows and loves us has created us in that way, which we do, then we have to have confidence in our capacity to know and love him. We must not be over-confident. There is no room for complacency. It is not a question of canonizing our every whim. There will be plenty of

mistakes. But a creator God of knowledge and love who programmes his knowing and loving creatures so that they are intrinsically incapable of knowing and loving him is an absurdity too gross to exist. We can be confident that we are not deceived.

2. Given, then, that we can have knowledge of God, there is the question of how that knowledge is made known to us. In the past we have heard of the divine inspiration of the Scriptures and of divinely revealed truths, which the Church teaches authoritatively. The implication—and sometimes it has been more express than that—has pointed to a distinct means by which divine truths have been communicated and made available to us and which, although of course addressed to us, operates independently of us. This distinct, independent, divine mode of communication has been seen as a guarantee of the status of the message. But we need to recognize that this knowledge is not given to us in some specially devised manner. The means of communication are familiar to us. I do not want to be misunderstood. The Scriptures are sacred. The Church does teach divinely revealed truths with authority. But the status of the teaching as divine does not depend upon a distinct method by which it is made known. On the contrary, were that the case, the Christian character of the message would be more readily compromised. In fact, however, what has been revealed to us has been unveiled in and through human means, human writing, a human community, a human being, Jesus of Nazareth. In him we believe that we find the perfect relationship between the human and the divine. The divine is perceived in and through the human.

Other principal themes will emerge as we proceed. We must begin by looking into the significance of Jesus of Nazareth for us.

2

Jesus of Nazareth

(i)

ONE Christmas two people stood gazing at a crib on the lawn outside a church. They remained silent for some time and then one turned to the other. 'That must be Adam and Eve,' he said. 'I wonder who the baby is.' The identity of that baby has been a cause of wonder for almost two thousand years. Born in Bethlehem, the son of Mary who was the wife of a carpenter called Joseph, he was named Jesus. After his birth the family settled in Nazareth. A name alone, however, cannot resolve our wonder. It supplies a clue, for 'Jesus' means 'saviour', but our curiosity is not satisfied so easily. Whose saviour is he? From what is he saving them?

What Christians believe about Jesus is totally implausible. We believe that if we are to grasp the meaning of all human existence, all that has passed and all that is to come as well as the present time, we must look to that child, born in an obscure corner of the Roman Empire almost two thousand years ago. We acknowledge the implausibility of the claim, but are compelled to affirm it nevertheless.

Our belief can be expressed quite simply. We believe that Jesus of Nazareth is God; we believe that he is a man; and we believe that he is one. In answer to the question, 'Is this man God?', we reply, 'Yes.' It is not that 'God is this man'; we can say much more about God than that. All the same, we affirm that this man is God. In him the divine and the human are so perfectly related that they are one. We do not hold the view that Jesus is God, disguised in human form, nor that he was so wonderful a man that a kind of poetic licence permits us to call him divine. We recognize the radical distinction between God and man. We know that what is human, what has been created, is not divine and that the divine is not human, creaturely, but, however inadequate it may be, we feel compelled to speak of Jesus in these terms. He is God; he is man; and he is one. To say so much may well seem very bewildering. I have stated these convictions starkly on purpose. Then, as we proceed, I hope gradually to explain them a little.

First of all, however, it is necessary to consider one prominent objection to this whole approach. There is a viewpoint, common amongst certain scholars today, that to speak of Jesus in these terms is no longer possible. To talk of him as both human and divine does not make sense. In earlier centuries it was different. Heaven was thought of as above the sky, and hell was below the earth. People lived in a three-tiered universe and to them the notion that God came down to earth from heaven came naturally enough. Or they were influenced by Greek myths which told stories of the gods coming and impregnating young girls who then gave birth to divine/human babies. Or else, it has been suggested, the early Christians wished to compete with such teaching and so supplied their own version. The general line of argument consistently presses the point that there was much in contemporary thought to make the teaching that Jesus was truly God and truly man readily accessible to the first Christians. According to this argument, such a way of speaking, described as mythological, is no longer available in our scientific, empirically minded world.

What bothers me about that position is its failure to fit the facts. It is important not to become so absorbed in detail that we lose sight of the picture as a whole. It could be that some modern scholarship on this question offers us a classic example of a wood being lost to view for the trees. When we consider what was happening in the early Church, we find almost constant controversy about who Jesus was. What we discover are various schools of thought. Some argued that Jesus was a man uniquely blessed by God. The favour shown him was so intense, so profound, that he was unlike any other man; he could perhaps be accounted as a kind of divine being; but he was not actually God. Others argued the opposite case. Jesus was God indeed who assumed human form, who appeared to be man, but he was not actually a man.

My purpose is not to give a detailed account of the various controversies about Jesus as the Christ which have raged in the Church over the centuries. The stories of Docetism and Modalism, Arianism and Apollinarianism, Nestorianism and Monophysitism have been told many times. They need not detain us. One point only we should observe. Seek out the common factor in these positions. It is plain to see. In one way or another, from one angle or another, none of them could accept that Jesus was truly God and truly man and one being. Each sought to tame that stark, implausible belief. Either they diminished the divinity, or they compromised the humanity, or else, acknowledging both, they denied the unity of the Christ. The difficulty is a feature of the early centuries as of our own. The people who tell us that these things cannot be said any longer, need to be reminded that they never could be said. It is a mistake to patronize the past. These controversies are falsified when drawn up as a dispute between those who won the argument and so installed themselves as orthodox and those who lost it and were dubbed as heretics, the orthodox being seen as thinking that it was easy to say God became man, while the heretics perceived the complexity of making such an affirmation and struggled with it. It was never an easy belief. The Church has insisted on it, because only thus could it give an account of its faith.

Consider one example. When Ignatius, the Bishop of Antioch, wrote to the Smyrneans on his way to martyrdom in Rome in 117, he spoke of Jesus as 'truly of David's line in his manhood, yet Son of God by the divine will and power'. Much no doubt could be said about these words. The thoughts that he had in mind when using them would have been rather different from the thoughts they stir in people today. Much has changed. World-views have altered, frames of reference have shifted, perspectives are different. Nevertheless at the root of it all, the conviction is the same. And every teacher, condemned as a heretic, resisted that conviction. He could not accept that, if Jesus was truly a man, he was also truly God; or that, if truly God, he was also truly a man; or that, if the God-man, he was one. Yet with increasing determination and in spite of the difficulties, the Church affirmed that what it believed could not be expressed satisfactorily unless the true divinity, humanity, and unity of the Christ were maintained. Nothing else would do.

How did such a belief come about? One important answer must affirm simply that that is the faith which the Gospels present to us. But it is proper that we should also try to search out the clues within them which may indicate how such a belief arose.

(ii)

In the 1960s great emphasis was placed on the humanity of Jesus. Many people thought that for far too long, although acknowledged, the manhood of Jesus had largely been ignored. In practice interest had been concentrated on his divinity. That was often true. Popular piety seemed to be absorbed with Jesus as God. However, that mood is now much changed. The humanity of Jesus takes precedence. Nobody doubts that Jesus was a man. Indeed there is something rather tired and dated about those books and articles which press on with warnings against neglecting the humanity of the Christ. You wonder whom they can be addressing.

We need to consider how Jesus came to be recognized as truly God and truly man. We cannot suppose that those who knew him, whether his apostles, disciples, and friends, or the scribes and Pharisees and those hostile to him, ever supposed he was not a man. A glance at the Gospels makes that plain. There is the famous instance when Peter confesses that Jesus is the Christ, the Son of the living God, but moments later is presented as blundering in on his plans, declaring that what Jesus has foreseen as happening to him in Jerusalem will not take place, and earning the rebuke: 'Get behind me, Satan.' That is hardly the action of a man who realizes he is addressing God incarnate. Again, during his trial, when the High Priest receives from Jesus an answer which could imply his divine nature, he rips his garments, for it is to him a blasphemy; the man standing before him cannot be God incarnate. What we now take for granted his contemporaries seem never to have doubted. The fascinating question concerns his divinity. How did the early Christians come to believe that Jesus was God? Some people would answer simply that Jesus himself told them. If he was God, as he was God, then he must have been aware of it himself and at the appropriate moment he passed on the information. That is straightforward enough, but it will not do.

Something must be said here which shocks many Christians when they meet it for the first time. They are shocked because they misunderstand it. What must be said is that nowhere in the Gospels does Jesus claim he is God in those words (Raymond E. Brown, *Jesus God and Man* (London, 1968), p. 30). That does not mean that there is not powerful evidence in the Gospels for the divinity of Jesus; there is. But Jesus himself never makes that claim. Various titles attributed to him or used by him may give the impression that he does, but closer study never leaves matters so simple. Let me give one example.

The titles 'Son of God' and 'Son of Man' have frequently been taken to refer to Jesus' divinity and humanity respectively. But 'Son of God' cannot be assumed to be referring to the divine; the Gospels show Jesus himself indicating that this can be a way of speaking about human beings, when he asks the Jews, 'Is it not written in your law, "I said, you are gods"?' (John 10: 34). And the title 'Son of Man', which has been the subject of extensive and detailed study, is considered to be a more exalted title than 'Son of God', for it points to a heavenly being, the Danielic Son of Man. This is complicated material; it is necessary to proceed with care.

To the man, Jesus, a Jew living two thousand years ago, the idea that he himself was God was completely unthinkable. There was only one God. Yahweh was God. No human being could conceivably identify himself with Yahweh. For Jesus to think of himself as God was beyond the scope of his understanding. This fact gives rise to two main questions. First, if Jesus was God, as we believe, then he must have known himself to be God in some way. It is unacceptable that Jesus should have known nothing of his own divinity. So the first question is: *in what way* did the man, Jesus of Nazareth, know himself to be God? And secondly, since from the early Christian times Jesus was believed to be divine, how did the first Christians come to recognize his divinity? I think it will help to take the questions in reverse order.

In the Gospels certain features, characteristic of Jesus, recur regularly: his relationship with God as Father, his sense of mission, and his authority. Now while the notion of God's Fatherhood was common amongst the Jews, the way in which Jesus spoke seems to have been unique. He seemed to be claiming an equality with God which had hitherto been unknown and which gave rise to charges of blasphemy against him. Yet it was this relationship with the Father, marked by an intimacy which his fellow Jews found scandalous, which inspired him and filled him with a sense of mission. He wanted only to do the Father's will. That was his constant motive. And in the fulfilment of that mission he spoke with authority. This authority is mentioned on various occasions, in particular early on in St Mark's Gospel. Those who heard his teaching were astonished, we are told, 'for he taught them as one who had authority' (Mark 1: 22). That is evidently not a matter of external force, but interior power. It is something that can be experienced only from someone who has a profound grasp on their subject. One day some years ago I was walking through the room in the house where I live where the television is kept. Elisabeth Schwarzkopf was giving a master class. I had not intended to stay, but I was gripped by the way in which she taught. It was authoritative. Here was someone who taught with authority. I understood the Gospel text afresh. It is a profoundly interior quality and, when it shows itself, it cannot be denied.

Leaving aside for the moment the effects on Jesus' disciples of seeing him risen from the dead, I suggest that this evidence would have had a powerful influence on the earliest Christians. They wished to understand more exactly who he was and they remembered his intimacy with God as Father, his sense of mission, and the authority with which he taught. As their understanding grew, they realized that they could give no satisfying account of these things unless they recognized him as divine. They had reflected deeply on their experience of Jesus and come to their conclusion about who he was.

This conclusion is all the more persuasive when we remember how adamantly monotheistic the Jewish Christians were. It went deeply against their most firmly held beliefs and instincts to affirm any kind of plurality in God. But their experience could not be set aside. Without yielding their commitment to the one God and without assembling any kind of systematic account to explain how it could be so, they acknowledged Jesus as divine. From that standpoint it may be easier to gauge the nature of Jesus' self-knowledge.

Every healthy person has a reasonable degree of self-knowledge. Those whose self-knowledge is seriously impaired are mentally sick. (This is something to which we will return later.) For Jesus to be in fact the one whom Christians declare him to be, he must have had a sufficient knowledge of who he was. But how clear was that knowledge, how articulate?

I am challenged: 'Who are you?' I give my name: 'Roderick Strange.' 'Thank you, but who are you?' my questioner persists and goes on persisting, even though I tell him about my family and friends, my work, my interests, my background, the story of my life. After each answer the question returns. My replies have all helped, they have all answered the question in part; but they have not exhausted it and I must admit that. In the end, I claim that I know who I am, but when every layer I can reach has been peeled away, there is still something more: the very depths of my identity are known to me, however imperfectly (for we need to grow constantly in self-knowledge), but it is beyond my powers to articulate that knowledge completely.

Did Jesus know he was God? Yes and no. Yes, because he must

have done. But it was a knowledge which he could never have expressed by simply saying to the Apostles one morning, 'Oh, yes, and there's another thing I wanted to mention, actually I am God.' He knew who he was, the well-loved Son of the Father, consumed with the desire to fulfil the Father's will and thereby speaking and acting with authority. As a Jew of his own time, he would have grasped his identity in terms of what he did, but the implications of his actions would not have enabled him to articulate his identity by saying simply that he was God. That was the knowledge deep within him, grasped but inexpressible.

Jesus was known as a man, recognized to be God. When we reflect further upon the relationship in him between the divine and the human, we are led naturally to ask another question: *how* is he both human and divine? Is there any help for the imagination here?

(iii)

The answer must be, not a great deal. We must beware of supposing we can portray the mechanics of mystery. Yet it may be possible to suggest a way of regarding how the divine and the human can be one, while each maintains its own integrity.

I think of musical harmony. In a sonata for violin and piano the music of both instruments is required to create the one piece of music. They do not produce a hybrid sound. Piano and violin produce their own music. They do not merge and become confused; each remains distinct, although inseparably joined to the other. Such an analogy may suggest to us how the divine and the human may come into relationship in the one Christ. Neither nature is compromised in the union, they remain unconfused while being united in the most perfect harmony.

Of course, the example is inadequate. The distinction which exists between two musical instruments and their subsequent relationship is quite straightforward compared to the utter distinction between what it is to be God and what it is to be man and their relationship. But analogies are to be employed for the help they can give, and not as though they were complete in every detail. Here it is a help to bring before us a distinction and a union in which the distinct parts retain their integrity while their union is profound. All the same, the perception of Jesus of Nazareth as truly God and truly a man can take place at a much deeper level.

Imagine two people visiting an art gallery, one an expert, the other untrained. They stand before a canvas in the section of the gallery devoted to surrealism. 'What does it mean?' says the untrained man as he gazes at the array of juxtaposed images. 'I cannot tell you,' his expert friend replies. 'Do you mean that it has no meaning?', asks the first. 'Oh, no,' comes the answer. 'It has meaning all right, but I cannot simply explain the picture to you so as to describe its meaning. You have to look at it carefully, allow it to make its impression on you, and then its meaning will come home to you, not at a level you can then express and analyse directly, but you will perceive it more deeply, within yourself.'

This is not a case of special pleading, but that way of grasping meaning in art can help us to understand how we can come to perceive the mysteries of faith. Like the various contrasting images on the canvas, there are some aspects we do see clearly. In the present context we know that Jesus was a man. We believe him also to have been God, because of the Scriptures and the faith of the Church. We know too that he was not some divided personality. Jesus was a single being. How someone can be truly God, truly a man, and an undivided person is something beyond us. It is not the parts themselves which make the greatest difficulty, but the problem of how the parts are related to one another. There is no simple answer. No straightforward explanation is available. But that does not mean that the whole is without meaning. A person needs to reflect deeply on the parts, allow them to have their effect, and in time he may come to grasp their meaning deep within himself. There can be no guarantee. Nothing is certain in this realm, for it is a matter not only of the mystery of Christ, as it is called, but also of the disposition of the one who approaches it. And that disposition is shaped crucially by the consequences of who Jesus is, God and man and one, and the significance of those consequences for us.

(iv)

At its deepest level the mystery of Christ seeks to tell us who we are, to show us, as I remarked at the beginning, the meaning of our existence. Christians do not believe that their humanity can be understood by reference to itself alone. Christianity is not a mere way of speaking about human values and dignity; it refers also to something other than the human, namely what is divine. And it believes that the truth about what it is to be human is uncovered only in relationship with the divine. This belief is not based on an abstraction. The man, Jesus of Nazareth, is the key, for he is God as well as man. In him the relationship between the human and the divine is discovered at its most intimate. And what is special about him is none the less significant for us. No man is an island, not even Jesus of Nazareth. His significance for us is understood more easily when we appreciate his solidarity with us.

The word, solidarity, may well have only political overtones nowadays; we think most readily of events in Poland and activity amongst trade-unionists in Britain. But the idea has deep roots. It speaks of people bound together because of what they have in common. Think of great events which stir us: Roger Bannister's four-minute mile, the conquest of Everest, the first man on the moon; indeed, consider the establishing of any new record. What is it that excites us? Most immediately, of course, we applaud the achievement of the individual. After that, however, and probably all unawares, I suggest that we are congratulating ourselves. We may never run a mile so swiftly, nor climb a great mountain, nor journey into space, but when such things have been done, they reveal human capacity: their achievement is our achievement. Our common humanity binds each one together, gives everyone a share in what each has accomplished, and reveals human capacity.

Jesus, we believe, was truly God and truly a man. Moreover, the relationship between the divine and the human in him was unique. But that uniqueness is not to be located in his humanity. In spectacular films, apparently ordinary motor cars perform extraordinary stunts; their engines have been souped up. If the example can be forgiven, we do not believe that Jesus is unique because his apparently ordinary human nature has been souped up by being brought into relationship with the divine. Were that so, the very basis of Christian teaching about the Christ and his significance for us would be destroyed. Jesus is saviour because his humanity is entirely at one with ours. His is not a specialized, unique model. He is one with us. His significance does not lie in the fact he is a special kind of man, that is, one who is divine, but in the fact that he is an ordinary kind of man who is also divine.

Jesus is a man, but also God, and so he reveals our capacity for intimacy with God. We will never be gods as he is God (most of us will never climb Everest either, nor walk on the moon), but he shows us that our human nature is capable of such an intimate relationship with God, the human is open to the divine. That is the vital issue and it is easily misunderstood. People expect to be able to observe the divine distinctly.

In general, it is natural to suppose that analysis will reveal the component parts of a single reality. If you investigate life in England between the two world wars, you can unravel the various factors and influences which were at work in society—political, social, economic, cultural, and many others besides. In reality, of course, they were intertwined in a very complex way; through analysis we can isolate them and discuss them one by one. And so people often imagine that, when they examine aspects of Christianity in which it is affirmed that the divine is present, they will be able to identify this godly aspect. To put the matter differently, they expect to be able to do so, and when they fail, they often conclude that this belief in a divine presence is mistaken. They are told that the Bible is God's word and expect to find a mark upon those documents explicable only by recourse to the divine. When their search fails, they decide that they have been misled. They hear the Church claiming to be the people of God and search out its distinctive character. They find a community as thoroughly patient of scrutiny by anthropologists, sociologists, historians, and the rest as any other human group and dismiss the claim as illusory. They learn about the sacraments as God's acts of love, but remember how widespread are rituals of cleansing, feeding, anointing, and so on, and classify them as conventions or even superstitions.

But if what I have been saying about Jesus is true, this general expectation is ill-founded in this instance. What 'God-made-man' tells us is that the human itself is the very dwelling-place of the divine. Although the human and the divine in themselves are utterly distinct from each other, the birth of the Word of God as man teaches us about their union through God becoming dependent upon man. The divine was not diminished by its human dwelling-place; the human was not compromised by the divine presence. They formed this perfectly harmonious unity. And we have to examine our human condition and try to discern the divine within it. It is a theme to which we will return on various occasions. The divine is always distinct from the human; it does not lose its integrity; but it is perceived by a profound scrutiny of the human.

This indwelling of the human by the divine is the basis of

Christian spirituality. As we live more and more in the Christ, so do we come to share more and more in the divine nature (2 Pet. 1: 4). It is heady teaching, but it has been a constant theme in Christian spiritual writing through the centuries. This life in the Christ is established by baptism. In practice it means that the lives of Christians must be stamped with that loving obedience which so distinguished his life.

3

The passion of Christ

(i)

CHRISTIANITY is a religion of salvation. For a long time men and women have recognized that their relationship with God must have been damaged in some calamitous way. Our Jewish forefathers in the faith told stories to give an account of what they perceived: Adam and Eve; Cain and Abel; Noah and the flood; the Tower of Babel. These stories tried to express the origin of sin. This sin was more than some particular wrongdoing; it tarnished the very roots of human existence. We speak of it as original sin. It is the source of all our sinning. Left to our own devices, we would be trapped by its power. Christians believe, however, that we have been saved from sin by Jesus of Nazareth who died for us on the cross. The cross has stood at the heart of Christianity. All the same, it is important not to see it in isolation. At times that has happened and the prominence of the cross has made people overlook the significance of the resurrection; that was seen as a happy aftermath to the main business which had taken place on Calvary on Good Friday. When in 1960 F. X. Durrwell published his book *The Resurrection*, he was praised by reviewers for recalling attention to this great event which had been so neglected. We will be turning to it later.

However, it is not enough simply to include the resurrection of Jesus with his passion. At times it might seem as though, in the final analysis, the last days of Jesus' life are all that truly matter. What went before was useful, some might say, the preaching, the account of daily events, the miracles can all be uplifting, but it was the dying and the rising that really counted.

I would want to challenge that view. Should we not say that the coming to birth, the life and public ministry of Jesus, his passion and death, resurrection and glorification are all of a piece? When we search out the inmost secret of Jesus of Nazareth, we find it in that intimate relationship with the Father which stirred in him the sense of mission which he carried out with such authority. You will notice at once that these are the raw elements of Jesus' self-knowledge. And within that relationship, Jesus' disposition was one of unfailing loving obedience. That was the motivating force of his birth and life, death and destiny, indeed of his entire existence. If we are to understand his passion and death, we must not restrict our view to those events alone, but we must see them as the expression of the love and obedience which entirely dominated everything he did.

(ii)

Once that has been established, we become aware that there have been many ways of understanding the cross of Jesus. They do not necessarily fit together, like pieces of a jigsaw forming a single picture. They are, rather, different ways of looking at the same reality.

One well-established view has pointed to the example which Jesus gives us in his suffering and death. It speaks of the trials that he endured and encourages us to cope with our difficulties by imitating him. If we suffer, then so did he. God's Son did not protect himself from the consequences of his humanity. He did not shirk the pain that is so prominent a part of the human condition. We can learn from him, and are stirred to a deep love for him through our recognition of what he has done for us.

A second approach has seen the cross of Jesus as the place where sin and death were joined in mortal conflict with love and life. The devil and the Christ struggled together for the fate of mankind. This way of speaking may seem highly mythological to people today, but, considered calmly, it expresses vividly the decisive clash between good and evil which Christians believe took place when Jesus of Nazareth, whose life was so dominated by love, was crucified by sinful men.

Thirdly, there is the theme of sacrifice and satisfaction. It has in the past often taken a powerfully forensic character. Human sin was seen to be an infinite fault, because it was an offence against the infinite God. Finite human beings were unable to make recompense for their failing, because as finite they lacked the resources to make the correspondingly infinite reparation. This terrible dilemma was resolved by Jesus. As truly a man, he was able to be the representative of the human race; as also truly God, the recompense he made to the Father was infinite and so cancelled man's debt. The scheme of this argument was neat and satisfying, not least because it underlined the need for Jesus to be truly human in order to be our representative, truly divine in order to accomplish our reconciliation with the Father. At the same time, it gives rise to objections. Its perception of God implies a sullen being, unwilling to forgive unless he receives his pound of flesh. What kind of forgiveness is that? It suggests little of the God of mercy and love. Then there is the further difficulty concerning the satisfaction which Jesus gives. His sacrifice of himself is efficacious because he is the sinless one. But in what sense is it just that someone who is without sin should suffer on behalf of sinners? Can the innocent ever make just satisfaction for the guilty? They cannot. If the sacrifice of Jesus, the ransom that he pays, is intended to be sensitive to God's justice, it does not easily fulfil that intention. Closer inspection seems to reveal a brutal God. On the other hand, what is valuable in this way of thinking is its sensitivity to God as the just one, its reluctance, in other words, to be so overwhelmed by the thought of his mercy and love that his justice evaporates. The way it describes the exercise of that justice may not be satisfying, but it would be unwise to lose hold of the basic insight: our loving and merciful God is also a just God.

When we ponder upon these ways of understanding the crucifixion of Jesus, it is important for us once again not to strain ourselves to make everything find a place in a neat pattern. The mystery of our redemption is not to be tapped so easily. We must rather turn to the method we used before. As we have allowed the humanity, divinity, and unity of the Christ to make their impression upon us in turn and may have found the meaning of that mystery as a whole at a level deeper than we can articulate, so here as well. We must reflect carefully on each aspect—the example of Jesus, the victory of the Christ over sin and death, and the sacrifice of the Saviour—so that we may come to grasp the significance for us of his death.

It may be helpful to say something more about sacrifice.

(iii)

Sacrifice has a prominent position in the Old Testament. From the earliest times it is depicted as having taken place. In the first pages of the Bible Cain and Abel are described as making offerings to the Lord (Gen. 4: 3–4). However, it was something which came to be heavily criticized by the prophets, in the psalms, and by Jesus himself. At a casual glance it may seem from the Bible that the time for sacrifice had passed, as though its practice was symptomatic of a more primitive phase in the Jewish people's history, which should give way when they reached greater maturity. But the casual glance is misleading.

Looking more carefully, we discover that sacrifices as such are not despised. It is not the offerings which are criticized, but the interior disposition of those who make them. A genuine sacrifice gives outward expression to an interior attitude. In rural communities, such as those of the early Israelites, often perhaps struggling for survival, it was a powerful act of faith to sacrifice the first-born lamb or the first-fruits of the crops. Here at spring or harvest-time were the essential means of life, but what came first was set aside as a sign of faith in God: his protection was being acknowledged as most necessary, as far more important than daily food. In these circumstances sacrifice evidently proclaimed an inner life grounded in faith. With greater security and increased prosperity, however, it is easy to understand how these attitudes to sacrifice might change. Sacrifice would be offered still, but the more settled and certain life could gradually have eroded the faith which it was supposed to express. And so the sacrifice could be offered without thought, a mindless act, becoming debased into superstition, something external lacking all links with the reality it was meant to signify. Is it any wonder that the prophet Amos will declare, '"I hate, I despise your feasts, and I take no delight in your solemn assemblies. Even though you offer me your burnt offerings and cereal offerings, I will not accept them, and the peace offerings of your fatted beasts I will not look upon . . . But let justice roll down like waters, and righteousness like an ever-flowing stream"' (Amos 5: 21–2, 24)? In the same way the psalmist announces: 'For thou hast no delight in sacrifice; were I to give a burnt offering, thou wouldst not be pleased. The sacrifice acceptable to God is a broken spirit; a broken and contrite heart, O God, thou wilt not despise' (Ps. 51: 16–17). And when Jesus himself is attacked by the Pharisees for mixing with tax collectors and sinners, he reminds them of the words of the prophet Hosea: 'I desire mercy, and not sacrifice' (Matt. 9: 13; Hos. 6: 6). He is emphasizing the same truth, namely that the quality of the inward life takes precedence over the outward fulfilment of the Law's demands. A sacrificial act is invested with worth through the interior disposition of the person who performs it.

Our regard for all sorts of people will lead us to make sacrifices for them; even at the slightest level an ordinary unselfishness will cause us to put ourselves out for complete strangers. Inconvenience is accepted because of generosity. Where the need is greater, the readiness to make sacrifices grows. In war or in times of crisis, people rally round to assist one another. When the bond is close, amongst friends, in families, between those in love, then we are prepared to suffer hardship all the more. With these thoughts freshly in mind, we can consider the passion of Jesus more easily.

Jesus died to save mankind from their sins. That is the simple statement of our belief. We think of the famous words: 'Greater love has no man than this, that a man lay down his life for his friends' (John 15: 13). Indeed we may recall what St Paul wrote to the Romans in a rather different context: 'While we were yet helpless, at the right time Christ died for the ungodly. Why, one will hardly die for a righteous man—though perhaps for a good man one will dare even to die.' Then he thrusts home his point: 'But God shows his love for us in that while we were yet sinners Christ died for us' (Rom. 5: 6–8). The greater the bond of love, the greater the readiness to make sacrifices. The greatest sacrifices are usually for those who are nearest and dearest to us. We can interpret John and Paul in two different ways, which nevertheless dovetail with each other. Paul emphasizes the outstanding quality of Jesus' love for us in his dying for our sake when we were in our sins, estranged, at a distance from him. His thought suggests that the ultimate sacrifice is being made not for the loved ones, but for those who were still hostile on account of their sinning. John's words about a life laid down for friends may remind us of the power of love which overcomes the barrier which our withdrawal into sin has set up. Our hostility is conquered by his love.

But there is the further consideration, the one with which this chapter began. We must not isolate Jesus' death. There is the danger that we can turn it into something merely external, a trick almost, or a cog in the mechanism of our redemption. We have to remember that there can be no true sacrifice without the interior disposition which gives it value. Jesus did not merely die for us, nor even rise again as well. His entire existence is of a piece. His fidelity to the Father's will bore fruit in a love and service of all men and women. On our behalf he came in loving obedience to the Father. That disposition shaped and guided everything he did. Perhaps we should acknowledge here that there is a sense in which the cross was unnecessary. We could have been saved from our sins if Jesus had died in his bed in extreme old age. The passion Jesus endured was not required arbitrarily, as though it were a vicious obstacle course which had to be completed before mankind could be redeemed. At the heart of human salvation, more significant even than the cross itself, was the fidelity of the Saviour to the will of the Father. To say that is not to reduce the meaning of the cross. Not at all. But it is to shift the perspective, allowing us to take Jesus' attitude towards ourselves to heart. Fidelity might seem a rather abstract notion, but not when it leads to torture and humiliating execution for our sake. Love and devotion were not only words on the Saviour's lips, but realities in his life to which he bore witness when he died on the cross. And the gospels tell us that the cross was inevitable.

(iv)

During his public ministry a time came when Jesus began to warn his companions and friends about the outcome of his mission. So in one place St Mark writes that 'he began to teach them that the Son of Man must suffer many things, and be rejected by the elders and the chief priests and the scribes, and be killed, and after three days rise again' (Mark 8: 31). We do not need to know precisely what was said. Some scholars would argue that the reference to the resurrection proves that it is a saying added with hindsight, because Jesus could not have known he would rise again; such knowledge would have breached the integrity of his human condition. I do not think we should allow ourselves to be side-tracked into that discussion. For the present purpose let us adopt a minimal position. It will be enough. Accordingly it seems reasonable to suppose that Jesus, like many another inspiring leader—to put it no higher than that—could foresee the consequences of what he was doing. He realized that his preaching was going to bring him into increasingly sharper conflict with the established Jewish authorities and he knew that they had the power to destroy him. (Perhaps the origin of Mark's reference to his rising again was a statement of his conviction that his mission would ultimately prevail, which is to say that the Father's will would be done.) The question for us involves rather the inevitability of what took place: the Son of Man *must* suffer, *must* be rejected, *must* be killed. Why did these things *have to* happen? Are we back once more with the idea of a brutal God, insistent that the Son should endure extreme human suffering before any reconciliation between mankind and himself could take place? Was the passion an obstacle course after all?

At this point we are moving from theory to try to realize what actually took place. Why Jesus died is a question which calls out for an answer in immediate and concrete terms, not merely on the level of abstraction. And here it may be instructive to recall a striking passage from Plato's *Republic*, which speaks of the truly just man, who seems, however, to be unjust. The argument here is very subtle and as such not relevant to our present concern. What matters is the implication that the just man seems unjust, because others fail to recognize his justice; in fact, his is a justice so pure, so exceptional, that his contemporaries are baffled; they can think of him only as wicked. What would become of such a person? We are told that he 'will be scourged, tortured, and imprisoned, his eyes will be put out, and after enduring every humiliation he will be crucified' (*Republic*, Book II, 361e–362a). This passage supplies a clue to our question.

The inevitability of the Son's suffering cannot be presented as a consequence of divine malice. It occurred because of the unavoidable conflict—which Plato had recognized in principle centuries earlier —between his perfection and the sinful condition of mankind. In theory we praise and applaud what is good. Practice is different. It is a commonplace to observe that saints are not easy to live with. No doubt that is true in the first place because, however holy they may be, they are not perfect yet and so they will do things which we may reasonably oppose; but in the second place, if we are honest, we will usually recognize in our criticism a deficiency in ourselves—our weaknesses are exposed by the quality of their lives and we instinctively try to protect ourselves.

When Jesus came, preaching the good news of the Kingdom of Heaven, commanding service and love, proclaiming peace and forgiveness, and calling for a change of heart and life, he was making radical demands which those established in authority felt obliged to resist. They had the power to destroy him. He knew that. Destroy him they did. They plotted his arrest, connived with the Roman authorities, and brought him to the cross. Obviously there are many historical and cultural matters here which are fascinating, but for which we do not have time or space. All the same, at one level we can affirm that these events took place as the natural outcome of Jesus' fidelity to his mission. They are not to be regarded as the fulfilment of some arbitrary design. No. Sin resisted love. The truly just man was crucified. And the cross stands, not as a mark of chance horror, but as the inevitable consequence in practice of the struggle between good and evil which was enacted decisively in the life, death, and resurrection of Jesus of Nazareth. But always it is his fidelity which lies at the heart of our redemption. He was sacrificed because he was faithful. This raises a question which cannot be avoided any longer.

(v)

How does the fidelity of Jesus save us? That is the vital question. The cross can be represented as a form of bargaining: mankind's sins are forgiven because of the virtue of this man's suffering. But to give so crucial a place to his interior disposition may seem to place us again at a distance from him. How can we share in what is interior? Are we, therefore, able only to learn from him, to imitate him? Have we simply to try to follow his example and be faithful too? Does this emphasis on his fidelity limit what is available to us to something extrinsic? Are we left with a Saviour who is our moral exemplar and no more? If so, we would be reduced to an arid nineteenth-century form of Protestantism.

In one way the answer to this question is being given gradually throughout this entire book. For the moment, however, it will be useful to give some pointers.

First of all, fidelity is not a particular or definable gift or talent like a good singing voice or the ability to bat against top-class bowling. It is a quality of a life. It affects everything a person is and does. Like justice or generosity or love, it defines us. It conditions who we are. Consequently, to be faithful as Jesus is faithful is not a matter of imitation, which would be external only, but of sharing in his life. As such it is precisely something interior. Our lives come to be shaped like his. There is a bond which unites us, a bond which takes root in and so grows out of our common humanity. You will remember that the significance of Jesus for us lies, not in his being a unique specimen of humanity in a way that separates him from us, but in his being an ordinary kind of man who is also divine. Our membership of the Church, our reception of the sacraments, and our practice of the virtues of faith, hope, and charity all contribute to our sharing in his life. They bring the bond to maturity. And it goes without saying that to share that bond is to live in fidelity to the will of the Father. But fidelity for the disciples takes its character from the master's fidelity. It was not an abstraction in the life of Jesus. If we wish to be his disciples, it will not be an abstraction in ours. We have been warned: 'If any man would come after me, let him deny himself and take up his cross and follow me' (Mark 8: 34).

(vi)

When I was first ordained a priest and people began to come to me with their problems, perhaps bereaved or because their marriage was collapsing, I used to be distressed at their predicaments. My distress was no doubt a sign of my own youth and good fortune. It saddened me that in so sunny a world these poor people should be unlucky enough to be coping with such agonizing hardships. I thought of them as the few, the exceptions.

Experience has made me wiser. I am still distressed by the sorrows I have to share, but for their own sake, not because I suppose them to be exceptional. It may be the case that there are some people who are never gravely wounded in their lives. If so, they are the exceptions. Most of us are wounded and indeed defeated by the wounds inflicted upon us. Some will have been bereaved. Others will have failed marriages or will have loved deeply and never been able to bring that love to fruition. There will be others who have never married at all and nurse always an unsettling doubt either that they do not have the capacity to give love or that they are themselves simply unlovable. Some people are trained for work, but denied the opportunities to use their skills. Some are unemployed. Some suffer from incurable illnesses. Some are handicapped. Some are victims of irreversibly tragic events beyond their control and become refugees or society's debris. And some again are overwhelmed by a sense of utter waste in their lives; it is nothing they can specify, but they know they are defeated. This list is not exhaustive.

In giving these examples I mean them all to have one thing in common: they have no bright side; they are all scenes of defeat. Of course, bereaved husbands or wives often marry again and happily; of course, those who are handicapped or victims of incurable diseases often give us examples of heroism which inspire us; of course, life's tragedies can be seen as challenges which we accept and meet. But it is wicked to suggest that it is good for people to die untimely young, or to be handicapped, or to find in their lives nothing but waste. Most people are wounded deeply and the wound does not heal. That wound can become the cross of Christ for them. It is not softened or removed by being accepted as such; the pain may be as bewildering as ever; but gradually a new aspect is being introduced. Let me explain what I mean.

One mistake which Christians can make all too easily is to consider Good Friday only from the vantage point of Easter Day. By doing so they miss a most valuable lesson. I believe that we need to see the passion and death of Jesus as his defeat. His sufferings and execution were not a game. There was no anticipation of resurrection to comfort him. His dying was not a charade. The Saviour of the world did not live through those terrible hours—confronted by the Jewish authorities, Pilate, Herod, brutal soldiers and mocking crowds—thinking to himself all the while, 'I know something you don't know.' He had come to Jerusalem with dark foreboding, but he must have been encouraged by the welcome he received from the crowds. Then, within days, his ministry was in ruins. He had not been looking for death. That was not his purpose. He came to preach the Kingdom, to call all men and women to lives renewed in service and love. He was about his Father's business. But sin rose up against love and brought him to a destiny which his fidelity to the Father's will made inescapable. One of the twelve apostles, Judas, had betrayed him; Peter, whom he had made their leader, had denied him three times; nine others had scattered; at the foot of the cross there remained only his mother Mary, his dearest friend John, son of Zebedee, and a few women of their company: of their pain he was himself the cause. Whatever he might once have envisaged as a happy outcome of his ministry must have been destroyed at that hour. He had been faithful, but his mission had ended in disaster. I do not believe it is possible to exaggerate the sense of failure which must have weighed Jesus down. And it is within this perspective that our defeats can acquire fresh meaning.

The spirituality which speaks of our being crucified with Christ can seem unreal in two ways. First, we fail to separate death from resurrection. The cross is always bathed in glory. The resurrection hides its horror. And so, secondly, as we tend always to see Calvary in the light of Easter, we invest it with a grandiose character and feel ashamed to mention in the same breath what we humbly see as our trials. We are mistaken on both counts. The cross can never be understood properly when it is only seen in the light of the resurrection. It has to be recognized for itself: it is the battlefield on which Jesus was defeated. And defeat is never glorious. Our modesty must not mislead us. It is not only the great martyrs who are defeated. When we are wounded deeply in life, when the ache of sorrow overwhelms us—bereavement, failed marriage, unrequited love, incurable disease, whatever it may be—then we discover ourselves to be utterly defeated, and we should turn to Calvary in our helplessness to remind ourselves that Jesus also was helpless and saw himself as a complete failure. Through our wounds we share his cross.

There were, however, two things which must have comforted him in his defeat: his conviction that, however unimaginable it might be, the Father's will was being done: 'Father, into thy hands I commit my spirit'; and his knowledge that his own fidelity had never wavered: 'It is consummated.' We need that conviction and that fidelity. They are the seed-bed of resurrection.

4

Risen from the dead to glory

(i)

ON 31 May 1944 a young RAF pilot flew his aircraft over occupied France and dropped equipment in preparation for the imminent invasion of Europe. He then turned north on an agreed course for the return to England. Over the coast of Holland his aeroplane was fatally attacked by night fighters. He circled the coast as best he could while the crew baled out. Three parachuted to the ground and spent the remainder of the war in prison camps. The other two crew members and the pilot himself parachuted into the sea and were drowned. They are buried in the war cemetery at Bergen-op-Zoom. It is an incident typical of war. I mention it here because of a remark that was made a day or so later. The pilot was my uncle and my mother has told me that when the news of his death reached home the family gathered together. My grandmother, his mother, was distraught with grief and kept on sobbing, 'We'll never get over it, we'll never get over it.' At last her husband turned to her. My grandfather was a Shetlander, as wise and gentle a man as you could wish to meet. 'No,' he said, 'we'll never get over it, but we will get used to it.' Ever since I first heard that remark, it has seemed to me to express, as it were in a nutshell, the reality of the Christian experience of death and resurrection.

To believe that Jesus rose from the dead is not a protection from all pain. He had to suffer and die before he could rise, and the kind of wounds which we have to endure when we share his cross are not healed instantaneously. The process is gradual. It can never be good that the Saviour was put to death by sinful men. Whatever good may have come from it, however vividly it has impressed upon us the power of his love for us, that brutal act itself can never be regarded as good. It is an abomination. And similarly, however we may respond to the wounds and defeats inflicted upon us, it can never be good that such tragedies occur. We never get over them. We never should get over them. To do so is to become insensitive to the reality of our own lives.

Yet we can get used to them. We learn to live with them. The situation does change in fact. How this happens can be a great puzzle. How does it come about that a person who is suffering from some profound misfortune which is not resolved nevertheless finds once more a joy in life and a sense of renewal which is far greater even than that experienced in past happy times? The fact of the experience is one to which many people bear witness. But how does it happen? We need to remember once more not to expect to be able to map out the mechanics of mystery. But something further can be added. I was discussing these matters with a student one day, puzzling over this shift from the wounded condition to a state of renewed life, and remarking that I could not describe how it took place. Suddenly he said to me, 'Isn't it something to do with the generosity of heart with which the wounding is accepted?' It was a comment which made sense of so much. We never get over the wounds we have received; our defeats are always defeats. However, if we allow ourselves to become completely absorbed in our own misfortune, if we cut ourselves off, then the defeat creates a crippling selfishness and is death-dealing indeed. But if we resist the whirlpool of selfishness and turn outwards, determined to show care and generosity and love where it is needed, if we keep our hearts open, then the transforming change can and will take place. Once again, the Saviour is our guide. His rising from the dead is no more to be regarded exclusively as an external act than his dying. As his death was a consequence of his fidelity, so was his resurrection.

To understand the cross of Jesus it is necessary, as we have seen, to look at it without considering the resurrection; but to understand it fully we have to realize its union with the resurrection. It was the old folly to invest all significance in the cross and to regard Easter as simply a happy aftermath. The Father would not accept mankind back into friendship without the Son dying on the cross. Once that had happened our reconciliation was made possible. 'Oh, and by the way,' people seemed to add, 'it was all right, because, although Jesus really did die, God raised him up again.' Obviously that will not do.

The interior disposition of loving obedience, the fidelity of Jesus to the will of the Father, lies at the heart of our salvation. It is a consistent message. His rising from the dead should not be seen only as miraculous, although in an obvious sense something beyond the normal pattern of natural events was taking place; it is more illuminating to place the emphasis elsewhere. Jesus was raised from the dead because he had been perfectly faithful to the will of his heavenly Father. In that setting we must affirm that his resurrection was the most natural and inevitable consequence of his fidelity. Our sharing in that fidelity becomes the springboard to our participation in the new life which Jesus offers. We are redeemed as his faithful people. With that securely in mind, we should go back and examine what this talk of resurrection means.

(ii)

In the gospel accounts of Jesus' appearances to his disciples after he had risen from the dead, two features are consistently present. On each occasion, almost without exception, something very direct, immediate, and physical occurs, but always there is something odd as well. Thus Mary Magdalene holds on to Jesus, although she had not at first recognized him, having mistaken him for the gardener (John 20: 11–18). The disciples on the road to Emmaus also failed to recognize him, but invited him in and he ate with them (Luke 24: 13–35). Thomas was invited to place his finger and hand into the wounded hands and side of the Christ; but the Lord had entered even though the doors were shut for fear of the Jews (John 20: 24–8). In St Matthew's Gospel the references are much slighter. Nevertheless here too we are told that when the women were running from the tomb with the news that Jesus had risen from the dead and he came to meet them, they 'took hold of his feet'; once again direct physical contact takes place. But when the disciples gather, as they had been asked to do, we learn that 'some doubted' (Matt. 28: 9, 17). There is this consistent combination of the physical and the odd. My favourite example comes in the final chapter of the Fourth Gospel. You will recall how Simon Peter had led them all off on an unsuccessful fishing expedition, how a figure had appeared on the shore at daybreak and encouraged the casting of the nets yet again, and how this had led to an immense catch. John and Peter and the others realize that the person who has been calling to them is Jesus and they hurry ashore. There he has prepared breakfast for them and they eat together. What could be more immediate? Then the Gospel tells us: 'Now none of the disciples dared ask him, "Who are you?" They knew it was the Lord' (John 21: 12). It is the word 'dared' which interests me. The disciples had recognized him. That is clear enough. Would it not have been natural to say that none of them

'*needed* to ask him, "Who are you?" They knew it was the Lord'?
Why does daring come into it, unless there was something about
Jesus which they could not explain or define?

What are we to make of this way of presenting the risen Lord in the
gospels? I suggest that both elements are trying to make a point. On
the one hand, the touching, the holding, the eating together are
meant to bring home to the reader the physical reality of what has
taken place. The risen Christ is not an illusion. The disciples'
experience is not born of wish-fulfilment. They are not victims of
mass hallucination. It is significant that the gospels never attempt to
cover up the way in which they behaved. We have noted previously
references earlier in the text to this resurrection, but his followers
were unprepared. When Jesus is arrested, they scatter, acting in a
cowardly and despicable manner. Their shameful actions are not
disguised or softened for a moment. Yet within a few weeks these
once timid men are out in the streets of Jerusalem and travelling
through the countryside preaching with confidence what they had
learnt from Jesus and proclaiming as the foundation-stone of their
preaching that God had raised from the dead the one who had been
crucified (e.g. Acts 2: 22–36; 3: 12–15). They are threatened with
punishment, imprisoned, flogged; in due course some are executed.
They do not waver. Even scholars who will not accept the reality of
the resurrection as an empirical event, but interpret it exclusively in
symbolic terms as a way of speaking about the continuance of the
spirit and aims of Jesus in the lives of his followers, will acknowledge
the difficulty of accounting for this rapid and deep-rooted change in
the disciples. They will often admit that the disciples must have
undergone an experience of incalculable power. That experience
arose from their being able to see and speak with, to touch and hold
and eat with, the friend who had been their leader and teacher and
whom, only a few days earlier, they had seen tortured and put to
death in a merciless manner as a common criminal. It arose from that
in part, but not altogether. The other element in their experience was
its oddness.

When the gospels speak of the disciples' failure to recognize the
risen Lord or of his appearance in their midst although the doors
were shut, they are trying to bring home to the reader a further
quality of his presence. To rise and to return from the dead are not
the same. Were someone who had died to return from the dead and
live once more, he would have to die a second time. But Jesus has

risen and death has no more power over him. He has passed through death to the new life.

This new condition can perhaps be brought into focus by raising an issue which Christians have considered from time to time. The matter can be put as a question. Suppose that in the upper room, besides the disciples, there had been some people who did not believe in Jesus, would he have been seen by both groups? The traditional answer, classically formulated by St Thomas Aquinas, has been to say no. The Christ was truly present, but visible only through the perceptiveness which faith bestows, *oculata fide* (*Summa Theologiae*, III, art. 55, q.2, ad 1). (The perceptiveness inherent in faith is something which will be raised later.) The physical element is evidently vital. It prevents us classifying the experience in exclusively spiritual terms. The oddness is important, for it prevents us classifying it only physically. They check and balance each other. They usher us into a new realm of life. When Jesus was raised from the dead, he revealed to us that new life. We should consider the significance of what he had done.

(iii)

At times that significance has been related closely to the empty tomb. Some people have seen in that tomb irrefutable evidence of the resurrection. It has had for them an immediacy which could not be denied. Others have seen the matter differently. They have argued that the state of the tomb was an irrelevance, because the belief which Christians hold is not concerned with whether or not the tomb was empty, but with a divine action. In this case they believe that God raised Jesus from the dead. They go further and argue the hypothesis that if the tomb of Jesus were to be discovered with his corpse still in it, their belief in his resurrection would be unaffected. The tomb, empty or full, is of no significance one way or the other. This is a delicate issue. It can arouse strong emotions and needs to be handled with care.

It is true, first of all, that our belief is in what God has done. It is not concerned primarily with the condition of the tomb in which Jesus was buried. Moreover, it is important to avoid identifying the teaching about the resurrection with the empty tomb. It could lead to ridiculous conclusions. To put it another way, we must take care that our position on the resurrection of Jesus does not depend upon there having been a body to bury. Jesus might have been burnt at the stake or taken to Rome and devoured by lions. There would have been nothing to bury. Would his resurrection have then been impossible? Of course not. God raised Jesus from the dead. He came to his death as a consequence of his perfect fidelity to the Father's will and in those circumstances death had no power over him. Its capacity for destruction had been overcome. It was now the beginning of his risen life. We are affirming our conviction about the true state of the human condition as revealed by Jesus of Nazareth. It is primarily a conviction which is spiritual.

What then is the status of the tomb? Is it an irrelevance after all? It is not. To say that our teaching on the resurrection and on the empty tomb are not identical is not to dismiss the significance of the tomb altogether, but to define it differently. The very fact that the question of the resurrection and of the tomb are not the same has puzzled scholars. They have been wondering why so much attention has been paid to the tomb. But there it remains, stubbornly, at an extremely early level of the tradition about the resurrection. It cannot be discarded. It bears witness to the fact that, whatever the profound spiritual meaning of Jesus' rising from the dead may be, at another more mundane level his body simply was not there. Setting aside the wild flight of fancy which explains the fact by suggesting the theft of the body by grave robbers, a coincidence of breath-taking proportions, we have to ask whether it was removed by Jesus' followers, who are the only other people under suspicion. If that were so, then Christianity has been built upon fraud from the outset. But that explanation will not do, not least because it is not required. As we have seen, at the primary level, the state of the tomb, empty or not, is not significant. It becomes significant at a secondary level as it earths the spiritual dimension in the physical. That is a pattern which we will discover in other contexts. However, if the empty tomb does not unveil for us the meaning of the resurrection, we have to seek that meaning elsewhere.

(iv)

Let me digress for a moment. In the ancient world mankind was aware of two haunting evils. One was death, the other corruption, the decay of our being, which follows from it. You will realize easily enough that corruption was perceived to be the greater of the two evils. Death was evil indeed, but principally because it was the threshold of subsequent dissolution. The pattern of human existence was seen as one of birth and life, death and decay.

The Christian gospel announced the breaking of that pattern. When Jesus rose from the dead a fresh pattern was proclaimed, no longer birth and life, death and decay, but birth and life, death and resurrection. The destiny of mankind had been changed through the invitation to share in the risen life of Christ. The corruption which had been so feared was undermined. Previously it had been understood as the inevitable consequence of death; now death was seen to lead to resurrection. This was the teaching of St Paul in his first letter to the Corinthians. 'This corruptible nature', he told them, 'must put on the incorruptible, and this mortal nature must put on immortality.' The evils of death and corruption are overcome through the resurrection. Not only corruption, but death also, for where is the victory of death as an evil when it is no longer the threshold of dissolution, but the gateway to risen life (see 1 Cor. 15: 53–6)?

The briefest thought will make us realize that this is not really a digression at all. The fears of the ancient world have reappeared in our society where Christian faith has begun to fade. As people today lose their belief in the promise of everlasting life, so does their fear of death grow strong. In the words of the Second Vatican Council, many are gripped by 'the dread of perpetual extinction' (*Gaudium et Spes*, 18). In these circumstances Christians must once again proclaim confidently their conviction that their gospel is indeed good news. Christ who was crucified has been raised from the dead. If we will die to sin, then we will be raised to new life in him. Here the meaning of the resurrection is unveiled. Sin and death and decay need hold no fear for us, for they will have no power over us. Once again we are back at the earlier point: if we are faithful, we will share in the life of Christ. It is a question of our interior disposition. We must turn away from the self-regard which destroys us to the generosity of heart which makes us at one with the Christ. His resurrection heralds the conquest over sin and death. That conquest is also an ascension into glory. It reveals to us our destiny. We are called to share in this new life. We will come to see that this life is not for the future only. For those who are baptized into Christ, it has begun already. For the moment, however, we should realize that this conquest over sin and death is also an ascension into glory.

(v)

The most obvious question to ask about the ascension is whether it is
to be distinguished from the resurrection. The event receives most
prominence in the writings of St Luke, but Luke is not helpful. In the
final chapter of his Gospel, after the discovery of the empty tomb,
we learn that 'that very day' two disciples were walking to Emmaus.
Their meeting with Jesus is described, ending in the meal where
bread is broken at which they recognize him. They rise 'that same
hour' and return to Jerusalem and the eleven. They hear about an
appearance to Simon and report their experience. During this
conversation—'as they were saying this'—Jesus stands amongst
them all, speaks to them, instructs them, commissions them as his
witnesses, leads them out to Bethany, and there is 'carried up into
heaven'. It is still Easter Sunday night. Turn now to the Acts of the
Apostles, also written by Luke. There we learn that Jesus 'presented
himself alive after his passion by many proofs, appearing to them
during forty days, and speaking of the Kingdom of God' (Acts 1: 3). So
when did the ascension take place? Is it another way of speaking
about the resurrection of Jesus, one unforgettable day, or was there
an experience distinct from that some time later?

    In the first place, we must not allow ourselves to be distracted by
the reference to forty days. When we muse over many biblical texts,
we may be struck by the frequency with which the number forty
occurs. The Israelites lived in the desert for forty years, Moses went
up Mount Sinai for forty days and forty nights, Elijah journeyed for
forty days and forty nights to Mount Horeb, and after his baptism
Jesus also retired into the wilderness for forty days. The number is
not to be taken literally. It refers to a special period of time. (There is
no obvious English equivalent, although we too do not always refer
to periods of time in a literal sense. When kept waiting for an
appointment, we may complain impatiently, 'I've been waiting here
for ages.') So when St Luke tells us that Jesus appeared amongst his
disciples for forty days, we should rather conclude that there was a
special period of time during which his followers had privileged
experiences of him after rising from the dead.

    In trying to appreciate this, I wonder whether we can be helped
imaginatively by other experiences which Christians have had.
Consider, for example, the experience of the visionary. Those of us
who have never had visions might suppose that those who do, have

them time and time again. But that is not necessarily the case. I think of Bernadette Soubirous in Lourdes. On 11 February 1858 this young girl had a vision of a lady who eventually declared that she was the immaculate Mother of God. She saw her on a further seventeen occasions until 16 July that same year, and never again. The visions occurred during a particular and limited period of time. Now I am not suggesting that Bernadette's visions are to be placed on the same footing as the disciples' experience of Jesus after he had risen from the dead, but her case may help us to appreciate that profound spiritual experiences can take place within a limited span of time.

It may be then that the disciples saw the Lord for a period of time which was brief enough, but not confined simply to Easter day. These meetings electrified them. Gradually the meaning of all that he had told them, all that he had done with them, the meaning of his passion and death began to break in on them. If you will tolerate a rather implausible simile, they were like someone who has spent years reading and enjoying plays without realizing that theatres exist. One day he is taken to see a play performed and it dawns on him little by little that what he has appreciated and treasured already is in fact the raw material for something far more magnificent which he has never previously even been able to imagine. Such was the effect of the risen Christ on his disciples. They did not understand everything, still less could they explain it perfectly to others, but the seeds which had been sown in the previous years began to ripen. By rising from the dead Jesus had not only conquered sin and death, he had revealed the glory of the Father in which he now shared. That is what his ascension proclaimed. And it was a glory not only for himself, but for the disciples as well: 'The glory which thou hast given me, I have given to them' (John 17: 22). The experience of Jesus, risen from the dead, unveiling his glory, stamped itself indelibly on them. But initial impressions, however powerful and vivid, fade. That is a feature of the human condition. When the privileged time ended, Jesus commanded the disciples to 'stay in the city, until you are clothed with power from on high' (Luke 24: 49). What had been started in them had to be continued in the Spirit.

(vi)

Hitherto, while we have been pondering upon the birth, the death, and the resurrection of Jesus we have been trying to notice also their significance for us. His birth affects us through our common humanity, his passion speaks to us of our deepest wounds which need to be healed, his resurrection and ascension supply that healing and proclaim our hope of glory. It is inspiring teaching, but how does it come about? In answer, we may begin by turning to an incident described in St John's Gospel. Jesus has gone up to Jerusalem for the Jewish feast of Tabernacles. He has gone privately to avoid the controversy that was growing about him, but from the middle of the feast he begins to teach. Then on the last day, the gospel states, he 'stood up and proclaimed, "If any one thirst, let him come to me and drink. He who believes in me, as scripture has said, 'Out of his heart shall flow rivers of living water.'"' These words are rich with meaning in the Fourth Gospel. 'Thirsting', 'drinking', 'believing', 'the flow of living waters' all point to a profound interior bond, binding those who believe into communion with Jesus, into a share in his life. Then the text adds: 'Now this he said about the Spirit, which those who believed in him were to receive; for as yet the Spirit had not been given, because Jesus was not yet glorified' (John 7: 37–9). In other words, the life in Christ which is offered to us is made available through his Spirit.

(vii)

From time to time Christians notice how rarely they pray to the Spirit. At Pentecost, we may sing, 'Come, Holy Spirit', but, in spite of the charismatic renewal movement, that is the exception rather than the rule. Occasionally we comment upon the 'forgotten' person of the Trinity. Why should this be so? Is it merely neglect or is there perhaps a deeper, more instructive explanation? Of course I am not going to suggest that it is a mistake to pray to the Spirit, but I think it unsurprising that we direct so few of our prayers to him. And it does not worry me.

It is a commonplace to remark upon the difficulty of defining spirit. It has been called a misty word, but an imposing reality. Without it talent, wealth, genius are of little use. (See Hugh Lavery, *Reflections* (Southend, 1978), p. 37.) We commonly associate spirit with wind or warmth. We may think of a yacht becalmed on a breathless sea. Then the wind rises, the sails fill, the yacht glides on. Or after a walk in winter we return home with numb fingers. 'My fingers have gone dead,' we remark to our companions. We hold up our hands close to the fire and, as the warmth spreads, we say, 'That's better, they're coming back to life.' Wind and fire, life and movement: these are qualities of spirit.

But spirit is more than that. It forges bonds. Writing to the Corinthians, St Paul tells them that 'no one can say "Jesus is Lord" except by the Holy Spirit' (1 Cor. 12: 3). That may puzzle us. Jesus or Joshua was not an uncommon name. The title 'Lord' is distinguished, but not applicable only to the man from Nazareth. Why should not many people have uttered those words, 'Jesus is Lord', without any connection with the Christian community or its Holy Spirit? The reason is made plain almost by formulating the question. Paul is not bothered about the physical capacity to utter those sounds. He is drawing the attention of his readers to a far deeper truth. The saying 'Jesus is Lord' had become for them a simple declaration of their faith. He is reminding them that no one can say those words and invest them with their true significance as an act of faith unless the Spirit of God dwells within him. These words acquire their character as an act of faith through the presence of the Spirit. Acts of faith are like passwords. To those outside the community they are meaningless; for those inside they supply their identity. This meaning comes through the Spirit who binds the separate individuals into union. The Spirit is not so much the one to whom we pray, but the one in whom we pray, yes, and live and move and have our being. He is the atmosphere we breathe. His stamp upon us makes us Christian, for this Spirit which we receive is Christ's Spirit.

People sometimes wonder whether the Spirit came to compensate for the absence of the Christ. But it is not so. The Spirit is present, not as second-best to the Saviour in his absence, but to make the Saviour present in a different way. We are on the fringe of mystery. Christ and the Spirit are not to be identified. It was not the Spirit who was crucified, nor Jesus who came down at Pentecost as tongues of flame. Nevertheless the Spirit is the Spirit of Jesus. He makes the Christ present and enables us to share his life. We are privileged people.

Again people will sometimes say to a priest, 'Oh, if only I could have seen Jesus. To know what he looked like, to have heard his voice, seen his gestures: how much more easily I could pray!' But such an attitude, understandable as it may be, receives no support in the New Testament. When Thomas wanted to see the wounded hands of Jesus and to put his finger in the mark of the nails, and place his hand in Jesus' side, his wish was granted. He had his way, but the implicit rebuke was overwhelming. The lesson was unequivocal. None of the disciples should look back to the days of the public ministry, when Jesus was in their midst, as a kind of golden age: 'Blessed are those who have not seen and yet believe' (John 20: 24–9). Nostalgia is never encouraged in these documents. And why not? Presumably because they did not see the public ministry in that light. Theirs was the golden age, the age of the Spirit. In the Spirit the Lord was present to them with a power and intimacy far greater than his physical presence alone could supply. Theirs was the golden age and that age of the Spirit is ours as well. The new life which Jesus came to establish and bestow he makes available to us through our union in the Spirit. We are only too well aware that it is something not yet accomplished. It is our *hope* of glory. We have to mature in Christ. Our perfection is to be achieved in and through our humanity. The power of the Spirit does not bypass that. To do so would be to contradict the meaning of the God-man's birth. But that same power dwells in us, assists us, and finally can transform us. Reconciled to the Father we are God's masterpiece, not the work of our own hands. He moulds us through his Church.

5

The Catholic Church

(i)

ON 11 October 1962 Pope John XXIII opened the Second Vatican Council. It sat every autumn for the next three years. Each session lasted about eight weeks. It was closed by Pope Paul VI on 8 December 1965. This Council was meant to renew the faith and life of the Catholic Church. Its influence has been vast and incalculable. However, even the fact that it happened should not be taken for granted. I was visiting St Peter's one afternoon early in 1966, before the seating for the two thousand and more bishops had been removed from the central nave of the basilica. I was approached by a woman in the portico. 'What has been going on?' she asked me in dazed and bewildered tones. 'I haven't been to Rome for twelve years. What are all those benches doing in St Peter's?' Four years of intense debate had obviously passed one person by without leaving a trace.

It is not easy to characterize briefly the mood and purpose of that debate, although what had happened was not unique. The book of Exodus describes the people of Israel during their journey through the desert to the promised land. A bare six weeks after their liberation from Egypt there were complaints: 'And the whole congregation of the people of Israel murmured against Moses and Aaron in the wilderness, and said to them, "Would that we had died by the hand of the Lord in the land of Egypt, when we sat by the fleshpots and ate bread to the full; for you have brought us out into this wilderness to kill this whole assembly with hunger"' (Exod. 16: 2–3). The reaction is familiar. Good things are promised and the promise is welcomed, but its realization is demanding and the demands are resisted. The experience of the people of Israel in the wilderness has in many ways foreshadowed that of Roman Catholics in more recent times. It is instructive to remember the backcloth against which the Council was called.

By the time Pius XII died in October 1958, many Catholics had come to recognize that everything was not well in their Church. Roman centralization was extreme and eventually the ageing Pope could manage no longer. Although greatly revered at large as a person, his leadership had ground to a halt. All the same, there were those who were comfortable with the old order, as well as those who were dissatisfied. The college of Cardinals was divided and so they elected the elderly patriarch of Venice, Cardinal Angelo Roncalli, as a caretaker Pope. The description has become famous, for he took care indeed, took care to inject into the Church an impulse for renewal. On 25 January 1959, only three months after his election, Pope John announced that he was going to hold an Ecumenical Council of the Church to help communicate the Christian gospel in a way that was more readily intelligible to contemporary society. Many people were delighted. Some were dismayed. But the path of renewal, so rich in promise, is also demanding. And there are, of course, those who now yearn for the Church as they knew it before Vatican II. They miss the security of that time, the clear sense of identity which they had had, the definite knowledge of right and wrong. The Council, on the other hand, had recognized that development could not be avoided. The Church was not so much a perfectly realized society as a pilgrim people. She is 'journeying in a foreign land away from her Lord . . . as an exile' (*Lumen Gentium*, 6). She is 'like a pilgrim in a foreign land'. She embraces 'sinners in her bosom' and 'is at the same time holy and always in need of being purified'. She 'incessantly pursues the path of penance and renewal' (*Lumen Gentium*, 8).

The Council's document on the Church, named after its opening words, *Lumen Gentium*, the Light of the Nations, is one of its major achievements. It cannot be summarized in a few sentences. Nevertheless, those brief quotations highlight for us one of its principal emphases. The Church was not to be seen in static terms, as an ideal already perfect, with its failings explained away by a strained distinction between the Church—the ideal institution—and the sinners who are its members. The Church is the people of God. They sin, they stand constantly in need of penance and renewal. But that need stirs them to grow in holiness. It is characteristic of the Council's teaching on the Church that it combines sinfulness and sanctity; it realizes that the two co-exist in fact; their relationship is a mark of the vitality which is necessary in a community which is going to remain alive.

(ii)

To understand better the shift in perspective which the Council sought in its teaching on the Church, it may be useful to recall more specifically the view that had been commonplace before it. In a book published in 1951 we find these words: 'Is it not food for thought that throughout all the changes of history . . . the Catholic Church has remained absolutely unchanged? In every age, in every civilization, in every country, she has remained just what she was when she first stepped out into the lanes of Jerusalem on that first Whitsun morning' (Francis Ripley, *This is the Faith* (Billinge: The Birchley Hall Press), p. 129). Let me try to unravel what is valuable and what is misleading in that quotation. The key is found in our understanding of Catholicism.

People often suppose that the Church is catholic because it is so large that it can be found in every corner of the globe. In other words, they understand catholic in a quantitative way: the Church is catholic, universal, because it is so big. But in fact Catholicism refers to a quality of the Church's life. The Church was catholic at Pentecost in the upper room when it consisted of nobody except the apostles and the mother of Jesus, and perhaps a small group of their friends who were gathered with them. Size was not important, but the quality of their lives was. 'Catholicism' means *wholeness*. The life of the Catholic Church amounts to more than the lives of individual Catholics. The Church is catholic when its members are trying to give a whole, complete response to God; and, provided that is being done, it remains catholic whether its membership consists of the entire population of the world or a tiny persecuted faithful remnant. It follows that being Catholic is independent of any particular age, civilization, country, society, or culture. The gospel does not require specialized conditions. On the contrary, history supplies us with too many examples of Christianity being impeded because there have been those who have insisted misguidedly upon presenting it in a particular and alien cultural guise. The handicaps imposed on Jesuit missions in China and India by a blinkered authority come readily to mind. The Catholic Church has not always remembered what Pope Benedict XIV expressed so succinctly: the desire that all might be catholics, not that all should become Latins (see Encyclical letter, *Allatae Sunt* (25 July 1755); quoted in H. de Lubac, *Catholicism* (London, 1950), p. 158). All this is implied by that quotation.

More problematical is the declaration that the Church has remained 'absolutely unchanged'. Of course Catholics believe that their Church is that original community of faith upon which the Spirit descended in the upper room. But she is not 'absolutely unchanged', she has not remained 'just what she was' on that first Whitsun morning. To some that will seem contradictory. In a debate a person might counter these remarks by arguing that if the Church is indeed the same Church, then nothing of essential importance can have changed, for a body which had changed in some essential way could not be regarded as having remained the same. The point seems to be logical and unanswerable. The answer, however, can be found by examining the nature of the enquiry more carefully.

We are trying to understand the Catholic Church, a community of people who have shared a common faith for almost two thousand years. They are not simply reducible to a logical entity. As a community they are embedded in history, historical. There is a crucial point here which must not be missed. When we describe someone or something as historical, we are not saying just that they exist in time: days, weeks, months, years pass them by. We are saying something about the way that they live: namely, the passage of time—those days, weeks, months, years—helps to form them, to give them their identity, to make them who they really are. Were the passage of time an irrelevance, an optional extra which human beings could take or leave, their essential nature would be different. The Church in the world is a living historical reality and, as such, it must develop. There is a famous saying which we could usefully remember here: 'In a higher world it is otherwise, but here below to live is to change and to be perfect is to have changed often' (J. H. Newman, *Essay on the Development of Christian Doctrine* (London, 1878), p. 48). The first quotation lacked the dimension expressed by the second.

We can consider various questions. What did that group in the upper room believe about God as three and God as one, about the presence of Christ in the eucharist, about the immaculate conception of the Virgin Mary? What was their understanding of redemption, their teaching on slavery and sexuality? Time permitting, could one of them have retired to a corner and composed the *Summa Theologiae*? These questions are not asked facetiously, but for a serious purpose. I hope that by setting them down it becomes possible to see that even matters at the very heart of Christian belief and life only come to be reached and known in the course of history. I shall try to say something more particular about these issues in the following chapter. For the moment we can notice that while we believe that the Church today is the same as the Church of grace established by the Holy Spirit at Pentecost, its identity is misrepresented when described as unchanged. The contrary is true. It has remained the same by changing. Its identity is preserved by development. There is no other way. And that point of view which has insisted upon the unchanged nature of the Church has caused a further handicap for Catholics.

To put it briefly, the ninety years or so between 1870, when the First Vatican Council came to an end, and 1962, when the Second was opened, stand out as a uniquely exceptional period in the life of the Church. There has been nothing else to compare with it. At Vatican I, the doctrine of papal infallibility was defined and that definition coincided with the Pope losing his power as a temporal ruler; in fact the Council closed prematurely as Italian troops prepared to enter Rome and put into operation plans for the unification of Italy. The Pope was confined like a prisoner in the Vatican. The loss of temporal power led to still greater emphasis on his spiritual authority. The degree of authority and control exercised from Rome became more and more intense. Other bishops were regarded—and often regarded themselves—as little more than delegates of the Pope in their own dioceses. Any idea of the significance of a local church had become very hazy indeed. The central Roman government maintained a firm hold. In the circumstances a well-defined, clear-cut position was adopted and taught. Many of the difficulties and subtleties which research was revealing, and of which scholars were becoming increasingly aware, were set on one side. At times, even to consider them was interpreted as evidence of weak faith and disloyalty. But that is another story. For the present we need to notice only that for almost a century the account of Catholic teaching with which people were familiar was fixed, definite, static, and unchanging to an exceptional degree. That is the point: it was exceptional, and viewed across the span of centuries its exceptional character is unmistakable. In the shorter term, however, many Catholics today, even those no more than forty years old, have been handicapped in their understanding of their faith because that exceptional account has been taken as the norm. Accordingly, they look at the complexities consequent upon the Council and see there the loss of a strong position, as it has been regarded. We need to realize that that strength was built upon too much simplicity and that if the Church is to continue to mature in faith, then it must come to terms with complexity and debate and accept within itself a range of diverse and complex positions. We should also realize that, in doing this, the Church is not embarking on an unchartered course, but returning to her native waters.

(iii)

A strong, clear teaching position is obviously a considerable advantage in practice. Subtleties can be ignored and inconsistencies ironed out. It is more easily understood and, moreover, can build up the image of the institution it supports. In the present case, it lent itself to the view of the Church as an ideal society, which, as God's Church, was already perfect. As I have mentioned earlier, when problems arose because of some undeniable failing, recourse was taken to the distinction between the Church as perfect and her members who were not: the song, not the singer. But the distinction can be made to bear too heavy a load, for the Church does not exist as an abstraction; it is made up of its members. And when this community with its claims to holiness and divine foundation is seen to be wanting, serious questions are being raised which are looking for an answer; they should not be dismissed with a glib distinction. Here are two such questions.

A year after the Second Vatican Council had closed, Charles Davis, who was a highly respected English Catholic theologian, announced that he was leaving the Catholic Church. His writings had done much to educate Catholic opinion in England about the issues with which the Council had been concerned, and his departure came as a real blow. Writing to explain his decision, Davis spoke of discovering a 'zone of untruth' in the Church. He added, 'I found discussion within the Church always limited by reasons of expediency, by consideration of what the present regime would take' (Charles Davis, *A Question of Conscience* (London, 1967), p. 75). More recently, details of the Vatican's financial affairs have received widespread attention, not only in the Press, but also in law. This book is not the place to comment on those issues in particular. I raise specific examples, however, because they can illustrate our problem. Davis was asking a question about the moral integrity of those who teach in the Catholic Church and so also a question about the truthfulness of their message. A gospel qualified by expediency is not the Christian gospel at all. The affairs of the Banco Ambrosiano touch a different sphere, but, if substantiated, would indicate a corruption which we would wish to regard as unthinkable in the life of the Church.

Let us assume the worst. It is a dispiriting prospect, but for the sake of argument let us assume that Davis's accusations are well founded and that the allegations concerning the Banco Ambrosiano are proven. Many books could be written about these matters; some have been written already. I raise the issue for one reason.

The Catholic Church claims to have been founded by Jesus Christ. It is one, holy, catholic, and apostolic. None of those marks, as we have seen already when considering the nature of Catholicism, is simply an external feature; they are all interior qualities. The unity of the Church is not in the first instance a claim to exclusivity or a comment on the state of the ecumenical movement; it derives from the Church's origin in the one God; conflicts between the various Christian traditions tarnish it, but cannot destroy that inalienable mark. Holiness is not mere moral goodness, a matter of obedience to external precepts; it is first of all a share in the life of the one God from whom all holiness derives; holiness is godly. And as catholicity refers to the quality of wholeness, so apostolicity refers to a profound interior bond which binds the community to a share in the faith of those men on whom the Spirit descended at Pentecost. And so the question is defined more clearly: if the Church bears these magnificent qualities, how can it hedge on the truth and act immorally? We must look for an explanation in our understanding of the incarnate Lord.

Jesus of Nazareth, we believe, was truly God and truly man, but, you will remember, there is a constant tendency for us to emphasize either his humanity or his divinity to the detriment of the other. There is a similar tendency in studies of the Church. Too often in the past it is the human side which has been dismissed unthinkingly. The stress on the Church as a perfect society was based on the appeal to its divine foundation, while the human element was neglected. The divine song was applauded without reserve; the human choir, frequently singing out of tune, was usually ignored. And the consequences have been serious. Our sinning—the untruthfulness, the corruption, communal and personal sin—and those scandals which arise in the Church make the claim to being a divine community implausible at best. At worst they make it ridiculous. But it is not a matter for ridicule.

There are always those who are anxious to explain away the divine nature of the Church. They examine its history, analyse its political position, assess the sociological factors, keep in mind the psychological need that human beings have for the religious, and pronounce that the phenomenon of the Christian Church is completely explicable without recourse to any suggestion of a divine character. And, of course, in one sense they are correct. But they need to remember that the divine origin of the Church is not to be sought in some way distinct from its human character. It is as human as the man who founded it, but as he was also divine, so is his Church. But there remains the vital qualification. In him divinity and humanity are united in a relationship of perfect harmony. Not so the Church. Christians have still to strive for and grow into that share in his life which God has granted them in Christ, but which their sinning compromises and obscures. Here is that teaching expounded by Vatican II: the Church is 'at the same time holy and always in need of being purified' (*Lumen Gentium*, 8). It is not some ideal institution. It is a community of sinful people who have been given the means to become saints. And so we bear those magnificent qualities—one, holy, catholic and apostolic—while at times we still hedge on the truth and behave immorally.

(iv)

Christians must be realistic about the Church. They have to recognize its human side as well as its divine, and work tirelessly to purify what is imperfect and sinful. At the same time, this viewpoint can raise an acute question for some Catholics. Reared on a notion of the Church's perfection and their membership of an élite, they can be disillusioned when confronted by this need for renewal. After all, they may say, there is obvious sense in belonging to 'the one true Church', but little point in putting up with the obligation to go to mass on Sundays and holydays, the laws on fasting and abstinence, prohibitions on birth control and divorce, and everything else, if this Catholicism is simply one Christian tradition amongst many. Some it may suit. Let them have it. The rest of us will choose according to our taste. Such can be the reaction.

There can be various reasons for it. One is plainly negative and indicated above. A kind of religious snobbery, the desire to belong to a select and exclusive group, can reconcile some people to quite severe demands. However, when that exclusiveness is breached and the weaknesses within the Church become evident, such demands may be judged excessive. In these circumstances we need to remember that Jesus himself sought out the company of sinners. It was the Pharisees who complained. We must resist the temptation to reduce the Church to a gathering of the sinless. In the Church we are united by our common faith and the fact that we all sin.

Another reason why people are disillusioned with renewal is more positive. In simple terms, the teaching about the Church which most Catholics have received until very recently identified in an exclusive way the Church of Christ with the institution of the Roman Catholic Church. I can remember a Catholic lady coming up to me at a social gathering after an ecumenical service in Church Unity week many years ago now. She had been talking to another lady who was not a Catholic. With a slightly worried expression she said to me, 'We are the one true Church, aren't we, Father?' Her attitude was not exceptional. Non-Catholics were only saved by chance, because they were too ignorant to know any better.

It is not difficult to illustrate the weakness of this position. Consider the status of those preparing to become Catholics who have not yet, however, even been baptized. We call them catechumens. It has been traditional Catholic teaching that if such people were to die during their time of preparation, they would nevertheless be saved. Although unbaptized in fact, they could be regarded as having received baptism on the grounds of their intention to be baptized. They were baptized by desire. It is moreover evident, according to this view, that those who have received this baptism of desire have secured their salvation more decisively than those who have been baptized but without being members of the Roman Catholic Church. We can see at once what has happened here, for baptism of desire, whatever its merits, is not baptism in fact. Such a person is regarded as saved because of his relationship with the institution of the Catholic Church, whereas someone else, although validly baptized, may find his salvation at risk for want of that relationship. In this scheme the non-baptized catechumen has 'more right' to heaven than the baptized non-Catholic. You can see what a confused mess this position is. By identifying the Church of

Christ with the institution of the Roman Catholic Church, the bond with Rome has become more important than baptism itself. That cannot be right, for it is by baptism that we are brought to share in the saving death and resurrection of Christ.

During the Council this great truth was clearly reaffirmed once more: we are incorporated into the Church through baptism (see *Lumen Gentium*, 11). Our baptism joins us to Christ, brings us to share in his life, makes us members of the Church, and leads us to salvation. Our baptism lies, furthermore, at the basis of our search for Christian unity, for there is a bond between the baptized (see *Lumen Gentium*, 15). The ecumenical movement is trying to make manifest what baptism has already established within. And the Council added a reminder to Catholics who allow the bare fact of their membership of the Roman Catholic Church to make them complacent. It warned: 'He is not saved, however, who, though he is part of the body of the Church, does not persevere in charity. He remains indeed in the bosom of the Church, but, as it were, only in a "bodily" manner and not "in his heart"' (*Lumen Gentium*, 14).

We have to be realistic about the Church. We must not attribute to it a perfection which, on the one hand, can fuel religious snobbery, and, on the other, rests on a deeply flawed viewpoint which identifies Christianity as Roman Catholicism and makes baptism marginal. In the wise words of the Orthodox theologian Evdokimov, 'We know where the Church is; it is not for us to judge and say where the Church is not' (see Christopher Butler, *The Theology of Vatican II*, rev. edn. (London, 1981), p. 119). When Christians of deep and sincere faith gather to celebrate the eucharist or spend themselves in the service of others, there plainly the Church is made present. But who will deny that presence whenever people—of whatever explicit faith or none—devote themselves in love to those in need?

To have said so much has brought us back to an earlier question. The acknowledgement of the constant need for renewal in the Church, of its imperfections and its scandals, and the recognition of sincere Christian faith in other Christian traditions, and, indeed, we should add, of authentic Christian living in those who apparently have no faith at all, may make us ask why we should bother to be Catholics. Why should we respond to the demands of this way? Why should we try to fulfil its obligations? What good reason can we give for being Catholic?

(v)

Everyone who comes to believe in the Christ is obliged to enter into the deepest and fullest possible relationship with him. To know Christ but to decide to keep him at arm's length, so to speak, as a fringe acquaintance, is a nonsense. The Catholic Church teaches that it has at its disposal the means for just that deep and full relationship. Let me mention some of those means at random. There is the bed-rock of faith in Christ himself. The bond with him is mystical, invisible, but also through the Church a visible bond. There are our sacred scriptures. There is the entire sacramental system which is the efficacious means for establishing us as members of the Church and bringing us to holiness. Within the Church there is the full priestly ministry, bishop, priest, and deacon; and here too the office of St Peter is carried on by his successor as Pope, the Bishop of Rome. There is our common faith, our common tradition of life, and richly varied spiritual traditions as well. And there is the very history of the community which binds us together. Stated like that, how bald it seems. When we ponder upon it, however, and take to heart what it implies, we discover a wealth of resources at our disposal for discovering the truth of God in Christ. There is more. We realize that no other Christian tradition is so richly endowed. Many may possess a large number of these aids, but no other has them all. (And, of course, in their view what they lack is interpreted as Catholic excess.) But Catholics believe that their account of the gospel is true as no other is true. It is not perfect. We can never exhaust the mysteries of God in this life, but it is true in a most privileged way.

Such claims that Catholicism is true can cause a wide range of reactions. Some people will be embarrassed, because it seems to conjure up once more the arrogance of talk about being 'the one true Church' in a way that was narrowly conceived and contemptuous of others. Such people are rightly aware of the excellence to be found outside the Catholic Church. In these circumstances it may help to remember that the truth is patient of expression in various ways, but that the different versions may not all be equally valid. Consider how a great play can be produced quite legitimately with different emphases. Was Gielgud's romantic Hamlet better than Olivier's athlete? How can we say? But occasionally we recognize something outstanding. In 1933 Sir Alec Guinness saw Ernest Milton as Hamlet. He has described its effect on him:

When I left the theatre I felt shattered. I returned to my bed-sitter in Bayswater, incapable of speaking for about twenty-four hours, knowing I had witnessed something I can only call both transcendental and very human, qualities which Ernest carried in the depth of his heart . . . He neither did Shakespeare a favour, as some actors manage to imply they do, nor kow-towed to him; he met him, in Hamlet, on mutual and loving ground. (*Blessings in Disguise* (London, 1985), p. 172)

The analogy is as ever imperfect, but it may help us to appreciate our conviction that what the Catholic Church teaches is an account of the gospel of Christ truer than any other. All versions may contain the truth, but they will not all be equally authentic.

To make such a claim seems to me unavoidable, because Christianity proclaims its gospel of salvation as the truth. We are saved by Christ, who is 'the truth' (John 14: 6); our discipleship brings us knowledge of the truth which will make us free (John 8: 31–2). That cannot happen if our teaching in fact is false or if its truthfulness is a matter of indifference. A sense of the difficulty with which truth is attained, is, of course, very proper and a necessary corrective to overconfidence, but it must not be taken too far. We must beware of being so overwhelmed by it as to make the truth unobtainable. Were that the case, it could have no bearing on our salvation, and the preaching of the gospel would be an irrelevance. But we teach what we teach because we believe it—for all its limitations and imperfections—to be the truest account that can now be given and we seek constantly to perfect it.

Should any Catholic read these words and feel rather pleased in a superior kind of way, there are two points that must be remembered. First, if this view is correct, a sense of superiority is completely out of place; it would be a sign of real foolishness. To be blessed with such knowledge of the truth is to be given awesome responsibilities. It is humbling to be called to bear witness to such a gift, not a cause for giddy conceit. And secondly, there is a warning, for it is evident that resources do not guarantee results. Brilliant original scholars, for example, have come from deprived backgrounds and privileged backgrounds have produced wastrels; and every Catholic should know that there are non-Catholics and non-Christians and non-believers the quality of whose lives can put us to shame. They have matured in patience and generosity, forgiveness and love, to a degree which compels admiration. There is no easy recipe for holiness, but once the way to intimacy with God

n truth and love has been discovered we must take full advantage of
t.

This knowledge of the truth is vital for salvation. Those who
vould turn their backs on it deliberately could not be saved. This
onclusion leaves untouched, however, the situation of those who
never hear the gospel preached or who hear it preached in so flawed a
vay that its truth remains hidden from them. This area is too vast to
be covered here, but we can notice that it highlights the supremacy
of conscience and it indicates the part to be played by believers as a
eaven.

For the good fortune of Christian life is not a treasure to be
guarded but a gift to be shared, whether directly by teaching or
indirectly through the quality of our lives. Belief in the gospel creates
esponsibilities to spread the word. Here we find the spur for
evangelization. There is no room for calculation, for attempting to
gain the most spiritual benefit for the least possible effort. In the
Lucan Gospel St John had a word with Jesus on one occasion.
Master,' he said, 'we saw a man casting out demons in your name,
nd we forbade him, because he does not follow with us.' He seems
o have been jealous for his own position. But Jesus replied, 'Do not
orbid him; for he that is not against you is for you' (Luke 9: 50).
What do you suppose John's reaction was to that answer? Do you
imagine that he felt aggrieved? Did he think to himself, 'Well, if that
man can at least cast out devils while remaining at home, why
houldn't I do the same? Why should I bother trekking round the
ountry after Jesus?' We know he did not. He was not calculating his
good fortune. He recognized that he had been greatly blessed. The
ove of Jesus embraced everyone but not everyone realized it. John
vas loved, and moreover he knew that he was loved. To be a
Catholic is not to be loved specially, for God loves us all; but it makes
possible the true knowledge of that love. There is no greater gift than
o be caught up in love and truth and to know it.

6

The sources of Christian self-knowledge

(i)

In 1981 four prominent Labour politicians, Roy Jenkins, David
Owen, Shirley Williams, and William Rodgers, announced that they
were leaving the Labour Party to found the Social Democratic Party
Although expected, their decision was controversial. It led to
acrimonious debate and the accusation of betrayal. In a nutshell they
replied that the accusation was inappropriate. It was not so much a
question of their leaving the Labour party, as Labour changing into a
party essentially different from the one to which they had belonged
and which they had served for so many years. They were founding
the SDP out of fidelity to long-held convictions. The Labour Party
was no longer what it claimed to be.

Moving closer to our subject, we can detect the same pattern of
debate at the time of the Reformation in the sixteenth century. The
reformers did not regard themselves as framing a new kind of
Christianity. On the contrary, they looked at the condition of the
Church and, as they judged it, discovered such abuses and corruption that they decided they could protect and preserve what was
original and authentic only by breaking away from Rome. Rome
had become anti-Christ; it was no longer what it claimed to be.

I give these two very different examples, one secular, the other
religious, because they both illustrate the recognition of the same
principle: the well-being or good health of a community, in a word
its identity, depends upon its true knowledge of itself. The charge
levelled by the SDP against the Labour Party and by the reformers
against Rome was that each had so far lost its knowledge of itself that
it was no longer what it believed itself to be. Now I am not
concerned here to settle either of those arguments, but the principle
at the heart of the disputes—and examples could be multiplied
endlessly—is embedded in the foundations of this chapter: being
your true self depends upon knowing yourself. It is not a question of
possessing perfect self-knowledge. Living communities, like

individuals, will grow in their knowledge of themselves throughout their existence. But there has to be a sufficient degree of self-knowledge. When even that sufficient degree is lacking in individuals, we say they are out of their minds, they are mentally ill; when it is absent from a community, the group may lose its identity altogether.

In this chapter we shall be examining the way the Church has acquired its self-knowledge and how it safeguards what it has learnt. As we do so, points which have already arisen will emerge again. When we considered the significance of Jesus, we noted that his identity was something which his disciples came to recognize. That perceptiveness will be significant here as well. Again, their recognition of his divinity was something they discovered in and through his humanity, not because he made a blunt announcement. We too must seek the divine in and through the human. And we must return to that vital matter of change and continuity. We have noticed that the Church remains the same by changing; but there is obviously more to the question than that. Some changes are essential, but others are destructive. Which are which?

These issues raise a series of important subjects: divine revelation and scripture; tradition and dogma; the nature of doctrinal development and authority. The material is vast. Let us approach it in the light of the Church acquiring and preserving its knowledge of itself.

(ii)

Catholic Christianity is a divinely revealed religion. It claims, in other words, that its distinctive beliefs are not simply the conclusions of human enquiry and as such on a par with the convictions which distinguish some other religious traditions, or philosophies, or even political parties for that matter. No. Catholics believe that their faith was given to them, or revealed, by God. At the same time, it should be made clear that we will need to examine the part played by human factors in that revelation, but for the present let us look more closely at the nature of revelation.

Until the nineteenth century in general and often more recently, Christians would explain that divine revelation was found in the Bible. The sacred scriptures contain the truths which God has wished to make known to his people. And this position was supported by pointing to the knowledge which was judged to be accessible only through the Bible: the account of creation, for example, or the history of the chosen people, or the teaching of Jesus; here was information which was available in no other way. Its exclusive nature confirmed its divine character. But time went by and scholarship developed. The history of the biblical region was investigated and its cultures examined, archaeologists made discoveries in ancient sites, and other documents from those hallowed centuries came to light. The consequences were a shock to many believers: the biblical account of creation bore no resemblance to the conclusions of science; the history of the chosen people was riddled with gaps and inconsistencies; even some of Jesus' most prized sayings were found not to be unique, as, for example, his teaching that the greatest commandment of the Law is the commandment to love God and our neighbour as ourselves. People were always looking for a way to express the vast and demanding Jewish Law in a single precept; there were other rabbis who had given the same reply. Whatever the kind of truth which the Bible contained, it was evidently not simply without error; something much more subtle was taking place. And its information could certainly not be classed as exclusive. The old faith, based on a belief in the Bible, seemed to have been destroyed. Different people have attributed its destruction to different reasons.

On the one hand, there are those for whom truth can only be expressed in literal terms. 'Don't beat about the bush,' they say, as they sit late into the night discussing original sin, the virgin birth, or the resurrection. 'Did it really happen?' Such people have to learn that not all accounts of the truth are presented in literal terms. You can tell the truth about something in words without composing, so to speak, a verbal photograph. Gerard Manley Hopkins's poem 'The Wreck of the Deutschland' is not untrue, although it would have been surprising had it replaced the report which appeared in *The Times*. The Bible contains many kinds of writing: narrative, legendary, mythological, poetic, proverbial, legal, prophetic, and so on. They are all telling the truth, but in the way appropriate to each. They are interested in the meaning of what has taken place. They probe behind the bare factual happenings to discover their significance. They arrange their material according to their primary concerns. Thus, for example in the Gospels, Mark lays stress on the revelation of Jesus as the Christ and gives prominence to the messianic secret, which unfolds that revelation so that it can be more accurately understood; Matthew organizes what he has to say around material gathered into five major discourses, perhaps for liturgical purposes: the Sermon on the Mount, the missionary discourse, the collection of parables about the kingdom of heaven, the discourse on the Church, and the discourse on the final times; Luke highlights the journey along the road to Jerusalem; and John's is a Gospel of signs. They are telling the same story, proclaiming the same truth, but in their own ways. And the deeper point is that the truth they express is not only often true in a non-literal way, but also in a way that can include mistakes. I am thinking of such examples as Job's denial of an afterlife. He knew it was a possibility, but decided against it (Job 14: 13–22). Or there is the view, powerfully present in much of the New Testament, that the second coming of the Christ is imminent. Errors of this kind merely bear witness to the human character of the documents. They do not compromise the essential truthfulness of these writings. Those unfamiliar with looking at what they read in these terms may find this difficult to accept at first. They may suspect that the truth is being sold short. If they persevere they will discover the opposite: it is being respected more fully and made available more completely than an exclusively literal approach could ever manage.

On the other hand, there have been those for whom the discovery that these documents are as patient of examination and as explicable as any other set of human writings has totally undermined their credentials as divine revelation. For them what is not exclusive is not divine. But implicit in that conclusion is an assumption which needs to be challenged. It is an assumption which we have come across before: if Jesus of Nazareth was truly a man, how could he be God? as the Church is scarred by scandal and so evidently explicable in human terms, what sense does it make to call it a divine community? And so here, if a thorough account of these documents can be given as human writing, what grounds can there be for classifying them as divine revelation? Each question rests on the assumption that our experience of the divine can only be authentic when it can be shown to be completely independent of the human. That assumption is radically unchristian. At the heart of Christianity there lies the conviction that the divine Son of God, who is by nature completely distinct from everything human and, indeed, everything created, willed to become dependent upon created human nature and was born a man, Jesus of Nazareth. His pattern guides us. A thorough human explanation cannot rule out the possibility of divine presence. We need to study the matter more carefully.

(iii)

Some readers, conscious of developments in these areas under consideration, may have had misgivings about the preceding discussion. They may have been worried by what they detected as a too-ready identification of divine revelation with the Bible. Something further needs to be said.

First of all, the revelation that has been given is not principally a message. It is not written information, or verbal knowledge, or a series of propositions. The revelation is not in words. The revelation is the Word, the Word of God who took flesh and dwelt amongst us, Jesus of Nazareth. The Christian revelation is the revelation of a person and it is received by our entering a personal relationship with him. His title as 'the Word of God' should remind us that he is himself God's definitive utterance to us. He is the one who makes the Father known to us in himself. He is the Way, the Truth, and the Life. There is no source of information superior to him. What he reveals is the whole truth about himself, about God, and about our salvation. No one else can add to what he reveals. He exhausts God's truth. Accordingly, we can understand more easily why the Church has taught that revelation came to an end with the death of the last apostle. Embedded in that teaching is the recognition that there is nothing to add to what Christ has revealed. Indeed some people have asked why we do not say that the time of revelation came to an end with the ascension of Jesus into heaven; but the Church's teaching bears witness to the unique relationship which the apostles enjoyed with him and is sensitive to what that relationship was able to unfold. Revelation is personal.

Once that position has been firmly established, something else should be added. That earlier viewpoint which had not grasped the personal character of revelation and saw it exclusively in propositional terms has had immense influence. But since it has been corrected, an opposite viewpoint has often been adopted. Accordingly there are those who have so taken the personal character of revelation to heart that they have been in danger of underestimating the part which words have to play. But words are not to be despised. They are one of the principal ways in which persons communicate, that is to say, in which they express themselves to one another. We have to be at ease with religious language. And so we can ask in what sense the scriptural documents are divine revelation.

(iv)

We have observed already that one of the principal objections to regarding the Bible as divinely revealed was based on the thorough account of its contents which could be given without any reference to a divine agent. But we have seen that its divine character does not depend upon our ability to isolate it from human activity. It is a matter of discerning the divine within the human. We need to remember a principal feature in the formation of the Bible. We have to ask how these particular books came to be selected for inclusion.

A common answer to that enquiry in the past has pointed to the authorship of the documents, Mosaic, Davidic, prophetic, and so on in the Old Testament, apostolic in the New. But literary criticism has played havoc with that reply. You do not have to be in thrall to the literary critics to acknowledge that different hands were at work on the Pentateuch, that King David did not compose all the psalms attributed to him, that parts of the prophecy of Isaiah were composed centuries apart, and that the second letter of Peter did not come from the apostle himself. Another answer has been couched much more in social or political terms. It has taken note of the interests of various communities who would wish for the prestige of inclusion: if Thessalonika's Pauline letters were to go in, then Corinth's would too. Leaving this potentially complex discussion on one side, we can look for our answer by concentrating on the community's knowledge of itself.

For many documents were composed, not only the ones that we have had preserved for us in the Bible. It is instructive to enquire why these were selected and the others discarded. Why include the Gospels of Matthew, Mark, Luke, and John, but not those of Peter or Philip or James? The second group are quite different from the first, it will be said. They are apocryphal, composed of passages which have obviously been invented, often by spurious Christian communities. It is plain that they are not authentic. Very good. So when is a document authentic and when is it not? The answer seems to be unavoidable. It is authentic when it articulates in a true way the faith of the Church. In other words, when the community knew what the document contained, they welcomed it because they found that it expressed for them their knowledge of themselves as Christian people. They were able to say about it, 'Yes, that is what we believe.' It captured for them the experience of the Christ, direct or indirect, which they had had and which lay for them, as for every Christian throughout the centuries, at the very foundations of their believing. They recognized their faith in these writings. The documents which are not genuine stand condemned through lack of recognition.

I realize that the process which selected documents for inclusion in the Bible was vastly more complex than this. Many influences were at work. All the same, at the deepest level, something of the nature which I have described must have been taking place. Deep down communities are Christian because of what God does in them. His Spirit dwells in them. Those who are responsive to the Spirit, we call believers. Believing through the power of his Spirit they seek to know and love God more completely, and it is his wish, as the God of knowledge and love, to be known and loved. The scene is set. And documents which were written by people caught up in the power of the Spirit, under the influence of the faith he generates (who they were precisely is of no importance), these documents were recognized and accepted at large by those who believed as expressing for them the faith which they had received from God. These writings articulated in a most special manner their knowledge of themselves. They expressed more than a human version of the faith. What had been written struck home so truly that it was recognized as God's word to them. These documents contain divine revelation.

(v)

The experience of Christ, received in faith, makes people his followers, members of the Church. Christians recognize that their experience has been expressed in a privileged manner in their scriptures; they feel compelled by what they discover there to acknowledge it as God's word, divine revelation. But on reflection we must acknowledge something else. This word of God is not uttered into a vacuum. It has a context. It is the possession of the community. It is God's word to us. To speak of the community here is to turn our attention to tradition.

Centuries of fruitless debate have distracted Catholics and Protestants as they have tried to thrash out the relationship between scripture and tradition. Protestants have claimed that all revelation is contained in the Bible and have been remorselessly critical of a Catholic position which seemed to propose a source of divine truth in the tradition of the Church and independent of the scriptures.

The debate has been fruitless—at least when argued from such starkly opposed positions—because it has failed to take account of the necessarily vital relationship between the two. The scriptures, as we have noticed, do not exist in a vacuum; the tradition, let it be stated without reserve, is not some additional source of information which cannot be checked and from which 'extra' saving truths can be plucked. Non-Catholics are sometimes disturbed by Catholic teaching. In particular they are critical of the status which the Catholic Church has accorded to some of its beliefs about the virgin Mary, her immaculate conception and her assumption into heaven. Such criticism is not to be dismissed out of hand. It places on Catholic theologians the obligation to work out and expound the scriptural basis for these doctrines. Beyond those controversial questions, however, all Christians must come to see that the teaching which their scriptures contain expresses their knowledge of themselves as a people sharing in the life of Christ, bound together by a common faith. The scriptures proclaim what the community is, they proclaim its identity. And that identity is also proclaimed, not simply in words, but also in action, in practice, day by day. In a word, its identity is proclaimed by the way it lives, by its life. That life is what we mean by tradition. The scriptures proclaim the tradition; the tradition finds expression in the scriptures. The two could scarcely be more intimately interrelated.

We can take the matter further. Divine revelation, we know, is not primarily the words of the Bible, but the person of Jesus Christ. All scripture bears witness to him. The Church's tradition does not mean simply its customary way of going about its affairs, but, first of all, it is its very life. It has life only when it shares in the life of Christ. All tradition bears witness to him. At their deepest level, both scripture and tradition proclaim Christ himself. They sing the same song in glorious harmony.

All the same, the teaching of tradition has frequently been cast in dogmatic form. We need to consider the nature of dogma as well.

(vi)

Dogma has a bad reputation. In the popular view it is undesirable to be dogmatic. The notion has a pejorative ring, associated with a fixed and narrow opinion which is dismissive of alternatives. Dogma comes from prejudice. But although that is the interpretation some people place on the Church's dogmatic teaching, the character of these dogmas is different.

Once again it is instructive to return to the Church's need for self-knowledge. Christian writers have often observed that the Church's beliefs were not cast originally in the form of dogmatic propositions. The point is obvious. The Bible is not like that at all. Dogmas, they have explained, were formulated because the teaching of the Church came under pressure. Groups would arise from time to time and elaborate interpretations which did not seem to be coherent with the common belief. Sometimes reflection showed that these positions were valid; they unveiled depths which had not been perceived before. On other occasions they were judged to be incompatible with the faith and were rejected as mistaken. What was happening in these investigations? The pressure of controversy was forcing Christians to understand what they believed more precisely. Those beliefs gave them their identity. So, to put the matter another way, they were engaged in a process of self-understanding. Their knowledge of themselves as Christians was becoming deeper and clearer. And they formulated their knowledge in carefully defined phrases or by giving privileged status to particular words in order to safeguard their self-understanding from error and to preserve it for future generations. The attempt by the Church to express its faith dogmatically was not an exercise in intellectual dictatorship, but a consequence of its growth in self-knowledge.

There are difficulties. Speaking briefly, how does the theory work in practice? To be effective, a dogmatic statement of Christian belief needs to be valid for all time. How can that be possible? Such an idea presupposes the kind of insensitivity to history of which we were so critical in our discussion of the Church. Truths are expressed at particular times in particular words by particular people living in particular cultures. But time passes, words change their meaning, human beings develop with each generation, and cultures rise and fall. How can a fourth-century, or even a sixteenth-century, statement of faith be of use to someone living now?

Here I think we have to consider the kind of effectiveness we are looking for from a dogmatic statement. To suppose that a statement from some far distant Council of the Church can be presented to Christians today without preamble and be at once a source of overwhelming illumination is obviously fantasy. Even when expounded with the necessary attention to its context, it is likely to seem remote; and how many people are going to have the time and possess the skills to examine it in that kind of detail? If dogma does have a part to play in the Church's life, it cannot be based upon an assumption that it is timeless. An analogy may help us to appreciate its role better.

An actor is preparing to write his autobiography. Throughout his research he gathers in a wide range of information: the people he has met; the places he has visited; the plays he has performed; the books he has read; there is much else besides. All this information will enable him to present his account of himself. As he proceeds, however, he becomes aware of something more particular. He realizes that a few of those people, places, plays, and books are special. They are special because of their influence upon him; they have made him the person he is. Anyone who does not appreciate the significance of his early love affair, his two years living in Paris, the production of an Anouilh play at Windsor, or his enthusiasm for *Wuthering Heights* can never understand him. Without them he would have been markedly different. They make up his essential heritage and he cannot deny them without being false to himself. His lover may be dead, the visit to Paris long past, the production at Windsor may have happened fifteen years ago, and it may be twice as long since he opened the Brontë novel, but their importance for him cannot be denied. I suspect that it is the same for us all. Although many people and experiences have formed us, the influence of a few has been indispensable. They have been essential in making us the people we are.

I find it helpful to understand the dogmas of the Church in these terms. They are not timeless truths in a sense that would abstract them from their particular historical and cultural setting. No human statement is. But they form part of our Christian inheritance. They are essential to us. They are the crucial insights into revealed truth which particular Christians have had at certain times, by which they have defined and safeguarded their beliefs and so their self-knowledge. As such they are indispensable. To discard them would be to deny a part of ourselves. Their immediate impact may have passed, just as memories of Paris may have faded, but the effect of the influence, and so their significance, remains.

I would like to add two further comments. We hold that these dogmatic formulations are important because they express an insight into revealed truth. That is a view which some people have questioned, because the historical process which leads to the definition of a dogma may seem indistinguishable from the one which leads to various theological opinions. If a dogma expresses a revealed truth, it might be argued, it should have a different pedigree from other religious views; whereas the real situation displays a variety of conflicting opinions from which one emerges in triumph as dogma, because, the sceptical would add, its proponents had more muscle. We are back with a familiar issue and it should be handled in the same way. Just as the biblical documents are patient of the same kind of analysis as any other piece of writing, but are recognized none the less as divine revelation, so the dogmas of the Church have a pedigree like any other point of view, but are perceived by the community to be articulating its divinely revealed faith. That is the pattern which we should by now expect.

My second comment concerns the language in which dogmas are expressed. This is a difficult matter. If American men can have trouble in Britain when they try to buy suspenders, what use can we have for statements of belief which are many centuries old? Yet language is not always as fickle as it is made out to be. It may develop rapidly and shift its meaning, but recognition of that phenomenon should not lure us into absurdity. The line of some arguments seems to suggest that language changes so swiftly that words have little value left at all. That is obviously nonsense. Words may change their meaning, but they do have meaning. And they serve a purpose. They are liberating. They do not imprison ideas; they set them free. Anyone who has to write, be they lowly student or best-selling author, will know what I mean. We have an idea, but we struggle to express it. We cannot find the right word. Then suddenly the word comes, the idea is born. It is not reduced by finding its word; it is released. Similarly the dogmas of the Church, the hallowed words and phrases enshrined in conciliar and papal declarations, have their value because they have captured aspects of our faith with particular exactitude. They usher us into further depths. Of course we now use those words no longer in the precise sense that they were used then.

But their meaning is accessible to scholars, even though only after painstaking study, and they in their turn have the responsibility of translating what they have learnt into ways that are intelligible to us. Moreover, it can be important for us to hang on to the meaning of some words. Some are irreplaceable. Imagine trying to describe someone in English with a single word which expresses his or her warmth of personality, easy manner, approachableness, feel for other people, and compassion. It cannot be done. Italians call such a person 'simpatico', although in fact the word is even richer and more nuanced than the English I have used can suggest. Now imagine further that in a hundred years' time, perhaps because of the flow of the work-force in the Common Market, Italians have come to mean by 'simpatico' no more than the English mean when they call someone sympathetic. There would then be occasions when Italians would describe a person as 'simpatico', but add, 'not in the sense that we use it now, but in the sense our grandparents used'. Some words can express an idea or a belief in a particularly precise manner. They deserve to have privileged status. Such are the dogmas of the Church.

There is, however, one great mistake which can be made. We have said that these dogmas articulate revealed truth in a way that is indispensable for us and so they help us to know who we are as Christians. All the same it would be a disaster if we ever supposed that what they express exhausts everything that can be or needs to be said. We will never be able to exhaust the truths of our faith. It is in the nature of such truths that they have depths unplumbed. There is always room for development.

(vii)

It has been suggested that talk about the development of revealed truth has disguised a dilemma. The matter seemed to be plain enough. Revelation had come to an end with the death of the last apostle, but what had been revealed was not known completely at a stroke. It took time for the Church to realize what it had received. Its understanding of its faith developed as time went by. That was the traditional view. Thus the dilemma. It was argued that if development meant no more than a making explicit of what had been known previously implicitly, then it was not truly development at all. It was merely a repetition of revelation, clearer perhaps, and better expressed, but not truly development. On the other hand, if development meant more than that, if it was not simply a tired repetition of revealed teaching, if it envisaged more than the movement from the implicit to the explicit, then it was introducing something genuinely novel into Christian thought. As a consequence, it could no longer be maintained that revelation had ended when the last apostle died. There was the dilemma: either a closed revelation and no development, or development and open revelation. But the very neatness of the dilemma should raise our suspicions. Christian life and belief are rarely so simple.

The revelation we have received, let it be said again, is not primarily propositional; it is personal. Personal experiences are complex. They affect us in many ways. Their development involves far more than the drawing out of a stream of logical propositions from their given premisses. A better analogy would be the production of a play. Attentive readers may have sensed already my enthusiasm for the theatre. One of my regrets is my failure to see Paul Schofield's *Lear* in 1962; but I remember the reviews. Critics spoke of the way his performance cast hitherto unseen light upon this classic tragic role. There is the significant point. No new edition had been discovered; no fresh scenes had been found; Shakespeare's play was Shakespeare's play. But the very wealth of that work had allowed something new to emerge from its production. The same might be said for Christian life as it develops. It gains its identity from revelation, but, as it lives out that vastly rich and complex revealed reality, truths, hitherto unseen, become visible. Only the dullest pedant would want to describe what is happening as repetitious, as the implicit becoming explicit. The revelation in view is unchanged; what has been given is complete. But lives lived under the sway of such a revelation continually uncover fresh insights, never perceived before. A closed revelation and new developments are not incompatible. The dilemma is a fabrication.

Let us turn to one final issue.

(viii)

A rich revelation gives rise to developments. So much we have seen. But we cannot assume that everything we may perceive as a consequence developing from this revelation is authentic. Some developments may be false and it is essential that the Church should be able to distinguish the false from the true. Various tests have been proposed. Cardinal Newman, who wrote at length on the subject, discussed seven in particular, but there has been fierce debate about their usefulness. They were in any case put forward more as indicators than watertight tests. For our purposes, we may prefer to adopt a different approach. It will be more in keeping with the theme of this chapter and, we may notice, was more favoured by Newman towards the end of his life. We must return once again to the principle with which we began: the good health of a community depends upon its true knowledge of itself. The community which knows itself will be able to detect what is alien to it.

We can concentrate upon three of the Church's primary activities. It offers worship. It teaches. And it governs, that is to say, there is an authority which must be exercised for the good order of the community. These three activities are interrelated and are able to hold one another in balance. Thus if worship within the Church were to predominate to excess, it would be liable to promote superstition, but wise teaching and good government would be a safeguard against such an abuse. Again, an undue emphasis on teaching could give rise to rationalism, but the prayer of the Church and well-judged authority would supply the antidote. And if authority itself were to slip out of control and become dictatorial, then the spiritual life of the community and the critical learning of its teachers would hold it in check. The three work on one another and ensure the good health of the Church at large. That health is the best protection against error. It must also be the major priority of those in the Church who exercise authority. The first of these is the Pope.

7

The papacy: its origins and teaching office

(i)

IN a famous statement Pope Paul VI once acknowledged that his own office was 'undoubtedly the gravest obstacle in the path of ecumenism'. The papacy creates controversy. It has inspired loyalty and loathing. What Catholics revere, Protestants suspect. A Catholic will rattle off Jesus' words to Simon Peter: 'You are Peter, and on this rock I will build my church' (Matt. 16: 18). What could be more decisive? he will ask. If so decisive, comes the reply, how did it happen that, when the Church met to deal with one of its earliest and most significant problems, namely its treatment of Gentile converts, Peter did not preside? He did not appear to have seen himself as special. He had abdicated in favour of James. The argument continues back and forth, becoming even more involved. Scriptural texts are compared, analysed, and interpreted; ancient sites in Rome are scrutinized by archaeologists; the development of the papacy and its influence are studied by historians. There seems to be no end.

This chapter can be no more than a sketch. Nevertheless, within those limits, I will try to present the main pieces of evidence which throw light on the origins of the papacy and outline the interpretation of events which they seem to me most naturally to suggest. After that we will consider the particularly controversial question of papal authority and its claim to be infallible.

(ii)

You are reading your newspaper at breakfast one morning and turn to a long article on the centre page. At a glance you see that one name crops up again and again. You conclude that that person must be of importance for the discussion. That is obvious enough. In the same way, when we look through the New Testament, we find that no one except Jesus is mentioned more often than Simon Peter. It is natural to conclude that this man has a special part to play. We may notice as well that sometimes it is not simply Peter who is in view. We come across incidents when Peter speaks, but clearly on behalf of all the apostles, and even one occasion when words addressed to Peter in particular are shortly afterwards addressed to the whole group of the Twelve: 'Whatever you bind on earth shall be bound in heaven, and whatever you loose on earth shall be loosed in heaven' (Matt. 16: 19; 18: 18). On reflection, however, these circumstances do not diminish Peter's importance; they serve rather to heighten it. He is the one who is picked out in order to bring events into sharper focus. This general frequency with which Peter is mentioned can act as a backcloth as we begin to consider three texts in particular.

For a long time Catholics have called attention to the exchange in St Matthew's Gospel between Jesus and Peter at Caesarea Philippi. Jesus asks the disciples who people believe him to be. They report various opinions. He asks their view and Simon Peter replies: 'You are the Christ, the Son of the living God.' Jesus acknowledges what he has said as something revealed to him by the Father and then in his turn tells Peter who he is: 'You are Peter, and on this rock I will build my church, and the powers of death shall not prevail against it. I will give you the keys of the Kingdom of heaven, and whatever you bind on earth shall be bound in heaven, and whatever you loose on earth shall be loosed in heaven' (Matt. 16: 13–19). You can perhaps imagine the quantities of ink which have been spilt over this text. I do not intend to add much more. Let us grant that the passage is not in its original form. The text in St Mark's Gospel has Peter recognizing Jesus as the Christ, but contains no parallel promise to himself. We can assume that St Matthew's account has been elaborated rather later. But then it may be the case that the fuller Matthaean version bears witness to the more developed understanding of Peter's role which was emerging in the early Church. As Christ is the foundation-stone, he is the rock—it is, of course, a play of words on his name: Peter means rock—the rock placed on that stone upon which the Church is to be built and against which evil shall not prevail. His possession of the keys is the mark of his authority, an authority whose exercise on earth has its effect in heaven. If Catholics have at times tried to make too much of these words, it is nevertheless a powerful text.

In St Luke's account of the last supper there is a moment when Jesus turns only to Peter. 'Simon, Simon,' he says, 'behold, Satan demanded to have you, that he might sift you like wheat, but I have prayed for you that your faith may not fail; and when you have turned again, strengthen your brethren' (Luke 22: 31–2). Here a promise of protection is being made and a duty indicated: Peter must strengthen the others.

Thirdly, at the end of the Fourth Gospel, by the Sea of Tiberias Jesus once more talks to Peter alone. He questions him three times about his love for him, questions which have been seen to correspond to Peter's triple denial during the passion. Deeply moved, Peter replies, 'Lord, you know everything, you know that I love you.' And Jesus gives him special responsibilities: 'Feed my lambs, feed my sheep' (John 21: 15–17).

No one should try to claim that these three texts, either individually or taken together, supply a proof for the Roman Catholic teaching on the papacy. Matters are rarely so simple. But it is noteworthy that these three documents describe Peter in the kind of conversation with Jesus which no one else has. It is a conversation in which particular duties are laid upon him: be the rock; strengthen the brethren; feed the lambs and sheep. Moreover, these texts have something else in common, something rather unexpected, but instructive.

On each occasion Peter is praised or promised protection or sheltered in love. He could scarcely be more privileged. And then, as much has been given him, much is asked of him: he receives his responsibilities. All that we have seen. So far so good. But what happens next? He is rebuked. In St Matthew's Gospel the conversation between Peter and Jesus is followed at once by Jesus foretelling his journey to Jerusalem, his sufferings, death, and resurrection. Peter protests, 'God forbid, Lord! This shall never happen to you.' Jesus then turns and says to Peter, 'Get behind me, Satan! You are a hindrance to me; for you are not on the side of God, but of men' (Matt. 16: 21–3). At the last supper, after receiving his commission, Peter seems elated. 'Lord,' he says to Jesus, 'I am ready to go with you to prison and to death.' Jesus brings him sharply down to earth: 'I tell you, Peter, the cock will not crow this day, until you three times deny that you know me' (Luke 22: 33–4). And even after their highly charged conversation by the Sea of Tiberias, when Peter notices the beloved disciple nearby, he cannot resist the intrusive question: 'Lord, what about this man?' And Jesus has to tell him that it is no concern of his: 'If it is my will that he remain until I come, what is that to you?' (John 21: 20–2). What can we learn from this juxtaposition of privilege and rebuke?

One reaction to the evident prominence of Peter in the New Testament is to cast it exclusively in personal terms. Accordingly it can be agreed that the fisherman Simon, son of John, named by Jesus Peter, was himself a significant figure in the early Church, but it is also argued that it was a unique significance, personal to him, and that once he had died his ministry came to an end. There is no 'Petrine function', as Roman Catholics claim, which is independent of Peter, essential to the Church, and a persistent feature of its life in each generation. But Catholics disagree. They accept that there was something personal about Peter's ministry which makes it indeed irreplaceable and unique, but they go on to argue that it cannot have been a ministry locked into its own time altogether. Christian generations do not exist as self-contained islands. It is the same Church, whether we are talking about the first century or the twentieth. Its essential character and activities persist. Peter's ministry to the Church is as much a part of its life now as it has always been. And these texts point to this truth for us. They indicate that the responsibilities laid on Peter were not completely bound up with him as a particular individual. His ministry cannot be identified with the private person who is still capable of making a fool of himself and needs to be rebuked. The rebukes are personal, while the ministry has a far wider scope. Throughout its history the Church will need its rock; it will need to be strengthened and it will need to be fed.

However, what are we to make of the evidence in the Acts of the Apostles? At first Peter is spokesman for the disciples, but when the controversy over the Gentile converts arises and a meeting is held in Jerusalem to decide whether they are to be obliged to fulfil the Jewish Law when they become Christians, he is not the one who presides. James has become the leader of the Church in Jerusalem and he conducts the business. Has Peter abdicated? He appears to be remarkably unaware of the responsibilities which Jesus has laid upon him. There are three points worth remembering.

First, we must beware of a tendency to see the early Church in terms familiar to us, but alien to itself. In particular we must be careful not to presume that there was a highly organized institution operative soon after Pentecost. All communities which last for any length of time develop some organization, but they do so only gradually when their origins are principally personal and spiritual. There is a whiff of anachronism about the argument which has demoted Peter because he is not in the chair.

And in fact Peter's contribution to the deliberations is presented as the crucial one. This is the second point to be remembered. A vigorous and unresolved debate has been taking place between Paul and Barnabas on the one side and believers of the Pharisee party on the other. It is continued in the presence of the apostles and elders. And then we are told that after much discussion Peter rises and speaks, pleading that the Gentiles should be accepted for their faith and the presence of the Holy Spirit in their midst and without any obligation as such to the Law of Moses. He wins a hearing for Paul and Barnabas who then give their evidence. The gathering is convinced by what they say and so James sums up and gives his judgement, but taking his cue from Peter: 'Brethren, listen to me. Symeon has related how God first visited the Gentiles, to take out of them a people for his name' (Acts 15: 1–21).

The third point is the most difficult. It concerns the kind of awareness Peter had of his responsibilities. If we allow for the fact that the Gospels do not give us an exact account of what took place, but have rather presented the significance of events and in doing so have supplied them with a clarity they did not originally possess, we must nevertheless hold that Peter had some real appreciation of the responsibilities with which Jesus charged him. If he did not, there is nothing to discuss. The account in Acts shows that he did. But we still have to ask about the kind of understanding he will have had of his ministry. Popular Catholic talk about Peter as the first Pope is vastly misleading. It seems to suggest that he will have had a developed, well-defined view of his role which he could have presented so explicitly that the Fathers of the First Vatican Council would have erupted into applause. And it is not only Catholics who are at fault here. Much non-Catholic discussion of the papacy, even of a scholarly kind, has dismissed papal claims because, it is said, they cannot be proved from scripture. That seems fair enough, until you realize that the request is for evidence of a Vatican-style bureaucracy well established from the Church's earliest days. That is absurd. I suggest that we do not need to suppose that Peter had a particularly developed or explicit understanding of his ministry. But he knew that he had to support and strengthen and nourish and he did so, not by planning an embryonic bureaucracy, but by going on

journeys, by travelling through the countryside, by preaching and
teaching in the footsteps of the Master, and—if the tradition is
correct—by dying on a cross upside down.

(iii)

Peter's death brought no sudden recognition for his office. The
evidence till the end of the second century occurs principally in four
documents. In 96 the Church of Corinth received a letter from
Clement who was the leader of the Christian community in Rome.
He showed concern for them in their difficulties and encouraged
them to repent of their sins, but the letter contains nothing which
might imply that Clement saw himself as having authority over
them. There were in any case strong links between Rome and
Corinth, which had been refounded as a Roman colony in 44 BC.
This letter is a consequence of those links. Then in 117, Ignatius, the
bishop of Antioch, was placed under arrest and brought to Rome to
be put to death. We would know very little about him, perhaps
nothing at all, had he not on his last journey written seven letters to
the churches and people whom he had met or whom he was about to
meet on his way. One letter was sent ahead of him to Rome. It shows
great love and respect for Rome, possibly even a distinct love and
respect, but here again the idea of Rome possessing some kind of
universal supremacy simply does not arise. Ignatius believed that all
the leaders of the churches had equal standing. This view is still
evident in the writings of Irenaeus in 178. Irenaeus is a highly
significant figure in the early Church. Born in the east, he became
bishop of Lyons. He brought east and west together in his own
person and wielded much influence. He recognized Rome as an
apostolic see, distinctively great because of its link with Peter and
Paul. He goes even further and teaches that union with Rome is a
norm for authentic belief. Nevertheless, he does not make any
explicit statement of Roman primacy, because he is confident that
any conflicts which may arise can be resolved by the fellowship or
communion of the bishops. Finally, back in Rome towards the end
of the second century, Victor, who was bishop, ordered synods to
settle the controversy over the date when Easter should be celebrated
each year. It is interesting that he went so far as to threaten with
excommunication those who disobeyed his ruling and failed to bring
their practice into line with his own. Here at last it may seem that a

successor of Peter is beginning to exercise his authority. But it is still more interesting to note that Irenaeus and various others blamed him for his severity and, as harmony prevailed, presumably the threat was withdrawn. Whatever authority Victor thought he possessed, it was plainly not recognized by the other bishops, nor does he appear to have had much confidence in it himself.

None of these documents supplies a proof of Roman supremacy. They show awareness of Rome's significance as an apostolic see, but there is no need to deny that other reasons will also have made it prominent. Political factors, its wealth, and especially its position as the capital of the empire will all have had their part to play. At the same time, the dominant motif is the connection with Peter. And belief in the divine institution of the papacy—the belief, in other words, that the papacy did not come about arbitrarily, but is the consequence of God's will for his Church—that belief does not depend upon the emergence of the papacy being devoid of all human factors. The familiar principle is at work. The divine reality of Christ and the Church, of revelation, of dogma, and now of the papacy (and, we could add, of the hierarchical order of the Church) is perceived and known in and through the conditions of normal human experience. It is a process which we should by now expect.

(iv)

The recognition of the Petrine function in the Church, which had not dawned by the end of the second century, was not going to emerge for a considerable time yet. There is not the space here for a detailed account of events, but we can point briefly to the position held by Irenaeus. His confidence that disputes which arose could be settled through the world-wide fellowship of the bishops had great influence. But the time came when his confidence proved to be misplaced. The Church had grown rapidly, particularly after the conversion of Constantine. A larger Church meant bishops who were more distant from each other, not merely physically, but doctrinally, as different and then conflicting schools of thought came into existence. Disputes had to be settled, but bishops were at times too partisan themselves to reach agreement. During the fourth century the pressure of controversy revealed the need for a still more fundamental way of resolving the difficulties of those in conflict. Little by little the contribution which could be made by the bishop of

Rome as the successor of Peter came to be recognized and accepted. By the middle of the fifth century, Bishop Leo of Rome is intervening decisively in the debate about Christ which came to a head at the Council of Chalcedon in 451. What had been adopted at first more as a practical expedient was gradually seen to have profound theological roots. Peter's responsibilities to support, strengthen, and nourish were recognized as forming a ministry necessary for the Church in every generation. Even so, there has never been a time when the papacy has been acknowledged explicitly everywhere; most obviously, it was never accepted formally by the Eastern Church.

The sequence of events which I have sketched is not presented as a proof of the papal office; but it is, I hope, worthy of consideration. Once we have realized that Peter's prominence in the scriptures is more than personal, for it points to a permanent ministry of which the Church will always have need, and once we accept that it was not established in an instant, complete and perfect, but, although a consequence of the divine will, had to emerge from the ebb and flow of human experience, we may agree that this scenario is very much what we would expect. However, a word of warning is also necessary. The demands of brevity may seem to have idealized what took place. It has not been ideal. Peter's less worthy successors have at times exploited their position for personal profit or political power. The history of the papacy has its own dark ages. How could it be otherwise? These men are only human. But it is a history of courage and self-sacrifice, of wisdom and saintliness as well. Time after time, this Petrine ministry has led the Church to understand itself more deeply and to safeguard its identity. Once again it is that concern for self-knowledge and identity which should guide our thoughts when we consider the authority of papal teaching and its claim to be infallible.

(v)

If the papacy itself has been a source of dispute, then its claim to teach infallibly has been even more controversial. People are often genuinely at a loss to know what it can mean. It appears to assert Roman freedom from error, when in fact Rome seems to make judgements which are about as right and as wrong as everybody else's. And they are puzzled to know the extent of infallibility. Does it cover every statement on any matter, however trivial, which is issued by the Vatican? Let us glance briefly at the more recent historical background.

Throughout the century before the definition of the dogma of papal infallibility by the First Vatican Council in 1870, there had been a tense struggle between Church and State in Europe. Various rulers wanted to have control over the papacy, while successive popes fought to preserve their independence. They rested their hopes on the Papal States. Hindsight enables us to see the irony of this strategy. It was their very responsibilities as temporal rulers which made them such pawns in the European power game, and the loss of the Papal States in 1870 has been the modern papacy's greatest good fortune. But the struggle intensified the need for independence which in 1870, with the Council in session and Italian troops approaching Rome, was cast in unassailable spiritual form as the doctrine of papal infallibility was defined. External pressures did not fatally compromise the integrity of what was done; the whole event supplies quite a neat illustration of the way in which God's truth can emerge in ambivalent human circumstances; but the circumstances have led in fact to a greater consciousness of and emphasis upon the infallibility of the Pope which have in their turn tended to heighten the controversy. Let us try to look at this teaching more calmly.

Once again, therefore, we return to the fact that the Church's preservation of its identity depends upon its having sufficient knowledge of itself. Were that knowledge to become insufficient, to be lacking in some vital way, its identity would be lost. Let us suppose, for example, that our belief in God as Father and Son and Spirit slipped from Trinitarianism into polytheism, or our understanding and celebration of the sacraments became mere superstition, and our moral behaviour reflected that erosion of belief. Would we any longer be the Church? Now the doctrine of papal infallibility does not cover every vague Vatican whim. It touches only the doctrines on faith and morals which have to be believed by the whole Church. In other words, it touches those matters of belief and conduct which are of universal significance. It is possible to say the same thing in a different way. The doctrine of infallibility is concerned only with those matters of faith and morals which are so fundamental that, were they lacking, the community's Christian identity would be seriously impaired; indeed it might not truly be a Christian community any more. It is a gift possessed by the Church to ensure that we never lose the essential self-knowledge which we need in order to be the Church.

Does this mean then, someone may ask, that infallible teaching gives us perfected beliefs and a corresponding perfect code of moral behaviour? It does not. The gift of infallibility acts negatively, that is to say, it does not so much enunciate what is true, as warn us against what is false. We can see that easily enough when we realize that with very few exceptions infallible statements have arisen in the course of a dispute as a corrective to erroneous teaching; and even the exceptions are never regarded as final statements—there will always be room for development. Self-knowledge must be sufficient; it will never in this life be complete.

As the Church is often described as a pilgrim nowadays, the image of a journey with infallibility as a guide may be instructive. If you live in Oxford and want to go to Stratford, you should drive northwards out of the city along the A34. It will take you all the way to your destination. We can call that the right road to Stratford. Alternatively, you could drive south out of the city, through Newbury and Winchester, arriving in Portsmouth. We can safely call that the wrong road to Stratford. Or you could drive west at first through Botley and then wend your way through innumerable villages until at long last you find yourself in Stratford. It could hardly be called the right road to use, but it is not the wrong one. Infallibility is like that. It does not promise the broad, uncluttered highway to a destination in perfect knowledge, but it guarantees you will not be helplessly lost. That guarantee should not be despised. I remember driving over two hundred miles one glorious summer day from Wallasey to Hemingford Grey near Cambridge. After a short while my little old car began to overheat and what I had looked forward to as a delightful journey became a wearisome ordeal. But I would have no hesitation at all in naming the worst moment. It happened when I thought (happily by mistake) that I had missed a turning on the motorway and was condemned to the wrong road for thirty miles. It is one thing to be making difficult progress in poor conditions, but much worse to be hopelessly wrong altogether. Infallibility is that gift within the Church which assures us that the worst will not happen. It should not surprise us that the Church possesses such a gift, for, unless it can guarantee that what it teaches on essential matters is at the very least not false, it is nothing. A community not only needs to know itself in fact; it must also have the capacity to understand itself and so come to true self-knowledge. The Church is infallible because it has that capacity. But how is this gift exercised?

(vi)

What we speak of as papal infallibility is not named as well as it might be. The gift is not the private possession of the Pope; it is rather, as our discussion has indicated, a gift which belongs within the Church, safeguarding its faith or, in other words, guaranteeing its essential self-knowledge. It is something exercised not only by the Pope, for the Pope is not the only authority in the Church. Every bishop has teaching responsibilities. When the whole body of bishops teaches in union with the Pope, then the highest authority in the Church is being exercised. This appreciation of the episcopal ministry has emerged more recently. It arises from an awareness that Peter's distinctive responsibilities did not isolate him from the other apostles. He was always one of them as well. And Peter's successor should not be isolated from the successors of the apostles; he is one of them and they are at one with him. This understanding of the episcopal office which gives attention to the bishops teaching as a single body is known as their collegiality and was another of the major themes to be expounded by the Second Vatican Council. It also complemented the teaching on the papacy presented by Vatican I.

So the gift belongs to the Church and is exercised by all the bishops together. But it is also a gift of the Pope's in a special manner. Holding Peter's office, he is the supreme pastor and so has final authority on earth in those essential matters of faith and morals which should be believed by the whole Church.

After giving a talk on the papacy on one occasion, I was criticized by one of my audience for not placing sufficient emphasis on the Pope's authority. It had not been my intention, but agreement with my critic was hampered by his firm adherence to the view that what the Pope taught authoritatively he had received directly from the Holy Spirit because he was the Pope. No other member of the Church had apparently any access to this information. Catholics must just obey unquestioningly whatever the Pope taught them. I may have streamlined, but I do not think I have caricatured, my critic's standpoint. I will try to express clearly—more clearly, I hope,

han I managed on that occasion—how papal authority should be
exercised.

We must remember, first of all, that every Christian who has not
deliberately severed his relationship with God has dwelling within
him the Holy Spirit. The Spirit gives us his gifts. He does not give
everyone every gift, but his presence, we may say, comes to fruition
in each one of us in the gifts that we need to fulfil our allotted tasks. St
Paul expressed it most simply to the Corinthians: 'Each has his own
special gift from God, one of one kind and one of another' (1 Cor. 7:
7). This presence of the Spirit and these gifts form us as members of
the Church and enable us to give it due service. The man who is Pope
is also blessed with gifts, for he too—only a foolish tendency to
isolate him from the rest of the community makes it necessary to
state the obvious—is a member of the Church. Like St Augustine he
will want to say, 'For you I am a bishop, with you I am a Christian.'
His special gift enables him to fulfil his office as the Church's
supreme pastor. More precisely, that means that he has the gift to
teach the Church authoritatively, especially in that limited, but vital
area which we were considering earlier, and his teaching has authority because it comes from the power of the Spirit. The Pope is
assisted by the Spirit when he teaches. However, the exercise of
authority has not been described completely when attention has been
paid only to the one in authority. Sports captains, army officers,
senior managers, union officials, prime ministers all have authority,
but its effectiveness depends upon the response given to their words.
Think of a shipwrecked sailor, radioing for help. He might as well be
whistling in the breeze if no one at all is tuned in to his frequency.
Much the same might be said about those in authority if their words
cannot evoke a response. We have to remember also the disposition
of those under authority.

We must realize, therefore, in the second place, that the same Holy
Spirit who is enabling the Pope to teach authoritatively is an
enlightening presence in every member of the Church. Authoritative papal teaching is not truly inaccessible to all other Christians.
This teaching is always an enunciation of revealed faith and as such it
is proclaiming something which is already known. Authoritative
word and living faith meet and recognize each other in these
pronouncements. There is another sense, of course, in which the
authoritative teaching was not known before. These pronouncements which articulate faith are creative. Putting it into words makes

a difference. Only the dullest pedant would claim, for instance, that lovers' first declaration of their love is unimportant, because 'they loved each other before and now have only put it into words'. Words matter. These authoritative statements clarify faith, develop it, and lead the faithful to depths which they have never plumbed before. Or perhaps we might say that Christians live in a landscape at dawn and such teaching gradually brings on the day. We must hold on to both senses: papal teaching tells us what we already know, while its very statement is what makes it known. And so this authority is not some kind of dictatorship over the Church, but a ministry of the Spirit which speaks to those who possess the Spirit and teaches them about their faith and their knowledge of themselves as God's people.

Within these reflections there can also be found the answer to the question people so often ask: could the Pope define error? The answer must be no, because the teaching which he gives is not a private invention of his own, but enunciates the faith of the Church. Authoritative word and living faith have to meet and recognize one another. The novelty can never be absolute. There is no new revelation. If, to put the case, a Pope in all the panoply of his office declared some teaching to be infallible which the faithful at large could not recognize as a part of their faith, that lack of recognition would be proof that what had been proclaimed was not infallible at all. This does not mean that the Church has a power of veto over papal teaching and can dismiss what it does not like; it is only the natural consequence when the teaching which is being given and the people who are being taught are formed by the same Spirit.

There is one other consequence. The Christian community must be characterized by a lively faith, a faith which is indeed on the alert to recognize the word of God which is authoritatively delivered to it. In other words, this understanding of the Pope's teaching office not only places a responsibility upon him to be a man whose spiritual life will make him a more apt instrument of the Spirit's power; it makes demands as well upon the quality of Christian life within the Church. That life is brought to birth and nurtured through the sacraments.

8

Sacramental life

(i)

CHRISTIANITY is a religion of love. In every Gospel Jesus insists upon its primacy to those who question him or put him to the test (see Mark 12: 28–34; Matt. 22: 34–40; Luke 10: 25–8; John 15: 12). It was a truth which the early Church took to heart. We may remember most readily St Paul's famous hymn to love which begins with the words, 'If I speak in the tongues of men and of angels, but have not love, I am a noisy gong or a clanging cymbal', and ends by affirming that of faith, hope, and love 'the greatest of these is love' (1 Cor. 13). But that is not an isolated passage. In the first Letter of John, we read: 'Beloved, let us love one another; for love is of God, and he who loves is born of God and knows God' (1 John 4: 7). And St Paul again, writing to the Romans, sums up simply: 'love is the fulfilling of the law' (Rom. 13: 10). There are many passages which could be quoted.

This love, which should bear fruit in the love we have for one another, finds its source and inspiration in God's love for us. The matter is put most plainly in that first Letter of John: 'In this is love, not that we loved God but that he loved us and sent his Son to be the expiation for our sins.' And he adds at once, 'Beloved, if God so loved us, we also ought to love one another' (1 John 4: 10–11). Our love for one another is a consequence of God's love for us, made manifest in the sending of his Son. That God loves us is not something which takes Christians by surprise, but we need to examine more carefully the way his love is expressed.

Many religious people understand divine love as a reality experienced on a spiritual plane as intense as it is pure, but for Catholic Christians the wonder of God's love is far greater than that. We believe that God has adapted his way of loving so that it corresponds perfectly to our whole condition.

Men and women are physical and spiritual beings. Our spirituality is indicated by our capacity for knowledge and love. Leaving aside bare factual matters, when we come to know a subject or a person truly, we find that we never know them completely. There is always more to learn. And when we love deeply, we discover the absurdity of ever saying we love so much that we are unable to love any more. There is no end to our loving. This unlimited capacity within us for knowledge and love is evidence of our spirituality. But our spirituality is bound up with our physical nature. We are limited beings, confined by our bodies and living in a particular place at a particular time. This must not be misunderstood. I am not implying that our bodies are a prison, caging up an otherwise unfettered spirit. I am trying to describe simply our real condition in which the physical and the spiritual are intrinsically and essentially related. Their relationship is creative.

It can be easy to assume that the spiritual aspect matters while the physical has only secondary importance. But when we consider our experience we find that the reality is different. We have noticed earlier, in our discussion of dogma, how finding the right word to express something which we know already, but cannot yet articulate, is a way of releasing what we understand and, indeed, of discovering it more deeply. Moreover, we can often observe how those who have become experts in some field of enquiry long to share what they have learnt with others. Some geniuses may be recluses. More usually they have an instinctive desire to teach—without, of course, necessarily possessing the talent to match the desire. Again, when we love, we long to give expression to the bond. Some families are undemonstrative, but more ordinarily love seeks expression. The lover who perceives love in an exclusively spiritual way and can find no reason for words of love, or for touching and kissing and sexual intercourse, is a lover incomplete. We can see here how well sexual intercourse is described as making love. When a man and a woman discover their profound union with each other and pledge themselves to each other, their bond most naturally expresses itself in sexual activity. They make love together. Their loving actions are creative of the love that binds them. (Conversely, there are times when couples believe a bond unites them, only to discover that their loving words have a hollow ring and their sexual intercourse has unveiled the superficiality of their relationship.) The human condition is one in which the physical and the spiritual are woven together inextricably.

God loves us, and the wonder is that his love takes account of our condition. He does not love us in some exclusively spiritual way. He loves us, as we have noticed a little earlier, 'by sending his Son' (1 John 4: 10). The love of God is made known to us in and through the birth, life, death, and resurrection of Jesus of Nazareth. Divine love is earthed in a particular human life. And this wondrous love which is so sensitive to the human condition, physical and spiritual, lies at the heart of the sacramental system. In fact it has become commonplace to speak of Christ as the sacrament and to recognize that all the other rites, baptism, eucharist, and the rest, which have been traditionally known as sacraments, are so only because they are extensions of Christ, manifestations of that divine love which God shows us in him in a way so considerate of human reality. God has shown his love by sending his Son. Every sacrament makes the Son present. They are acts of love which touch us as we are.

There are many people, however, who feel ill at ease when asked to consider the sacraments. Some non-Christians, particularly in the western world, may think of symbolism as alien to them and many who are Christians—and even devout in their practice—may feel insecure. Talk of a man, Jesus of Nazareth, or a human community, the Church, or of such characteristics as faith and hope and love all seems manageable. Those subjects are recognizably part of our experience. But this talk of sacraments can make us apprehensive. Sacraments can seem so technical. We wonder what we are really supposed to be believing.

Volumes have been written on the sacraments. They can indeed seem very intimidating. Accordingly it is all the more necessary to emphasize that here too our experience can help our understanding. On reflection we can discover that the words and actions which express our interior lives of knowledge and love are not external and distinct from those lives, but caught up in them. They are a part of the reality they symbolize; they can even be creative of it. Think once more of a pair of lovers. No one supposes that the words 'I love you', and the kiss which so often accompanies them, are to be identified as the very love itself; but by the same token only a fool would argue that the words and gestures which symbolize love are altogether distinct from it. Similarly, we believe that by baptism we come to new life in Christ. We celebrate the sacrament by pouring water and saying certain words. Water suggests cleanliness, purity, quenching of thirst, and many other things besides; the words proclaim the new life in the name of the Father and of the Son and of the Holy Spirit.

The new life is not to be identified with the washing and words, but they are its sign. The symbol brings the reality into effect. Symbol and reality intermingle. The symbol helps the reality to exist; the reality finds expression in the symbol. So those who feel apprehensive about this whole area of symbolism, of sacramental sign, can gain the confidence to move in it more freely by reflecting upon their own experience. While there are many highly sophisticated ideas which could be discussed, the issue fundamentally derives from something with which everyone is familiar. Our deepest reality is expressed and even in part created through symbols.

Closely associated with this awareness should be the recognition that symbolic acts are deeply personal. Initially that may seem improbable. A routine pattern of words and gestures repeated again and again may seem deadening. Children and teenagers will tell you that they find the mass boring. However, when we think about it a little harder, we may come to a different conclusion. Permanent spontaneity is a contradiction in terms. In some Christian traditions we have to be always on the alert in case we miss whatever may happen next. But a routine which is familiar is not going to give us those sudden surprises, and so it allows us to settle and to perceive the more profound implications of what is taking place. Established conventions assist communication between people. Generally speaking, we are uneasy when total strangers introduce themselves to us by hugging us warmly. We do not normally analyse our reaction, but I suspect it is not a question of being opposed to physical contact; rather we are resisting a situation in which a sign of intimacy has been adopted by someone completely unknown to us. Its meaning has been obscured. When those meanings are protected, when the familiar routine is preserved, we can then dig deep.

Thus far we may conclude that the sacramental system bears witness to God's love for us, a love which can be recognized all the more clearly from the fact that it respects our condition in which the spiritual and the physical are interwoven. Nor is the use of symbolism alien to us. On the contrary, by considering our experience we may perceive that it helps us to live our lives more profoundly. But there is one area in which symbols are common, namely superstition. People sometimes ask how sacrament and superstition can be distinguished. It will perhaps be useful to say a word on this question before we continue.

I heard once of a detective who only ever wore a hat when

nvestigating a murder. Previously he had never worn one at all, but
when assigned to his first murder case as a young detective, it was
aining so heavily that he dumped on his head an old hat which by
hance was lying in the back of his car. He solved the case and wore a
at for those enquiries ever after. That is a superstition. It was not
mportant. It did not matter, but that is what it was. You could say
hat wearing a hat became a ritual act when he was involved in a
nurder investigation, but the difference between this ritual and
acramental ritual is plain enough. The wearing of the hat had no real
elationship with nor bearing upon the investigation; but Christians
believe that sacramental acts bring into effect the reality they signify.
t is time to look at them in more detail.

(ii)

The Catholic tradition has taught that there are seven sacraments,
baptism and confirmation, the holy eucharist and penance, marriage
nd holy orders, and the sacrament of the sick, once known
generally as extreme unction. It has been customary to say that these
acraments were instituted by Christ. The grounds for that teaching
were sought in the scriptures. Some cases were more easily satisfied
han others. The conclusion of St Matthew's Gospel contains Jesus'
command to go and make disciples of all nations, 'baptizing them in
lie name of the Father and of the Son and of the Holy Spirit' (Matt.
8: 19). At the last supper Jesus is seen to celebrate the holy eucharist
or the first time (Matt. 26: 26–9; Mark 14: 22–5; Luke 22: 19–20;
Cor. 11: 23–5). Those examples seem straightforward. But then
people ask when the sacrament of penance was revealed to the
apostles by Jesus. A ready answer is given. In St John's Gospel Jesus
ells them, 'If you forgive the sins of any, they are forgiven; if you
etain the sins of any, they are retained' (John 20: 23). But the more
persistent enquirer will draw attention to the lack of any evidence for
he celebration of this sacrament until the third century and to its
nore frequent celebration only beginning with the practice of Irish
nonks in the sixth. The words in the Gospel are then interpreted as
eferring far more naturally to the administration of baptism by
which sins are forgiven and people reconciled to God. In the context
t is difficult to see how else they could have been understood. Or
again marriage is said to have been instituted at the wedding feast of
Cana. But that was the occasion when Jesus is presented in the

Fourth Gospel as performing the first of the great signs which in the Johannine account give his ministry its unity. Here water is changed into wine. The setting of a marriage feast has little significance at all.

What then can we say about the origins of these sacraments? It is unarguable that they were not instituted by Jesus in the direct way that has often been envisaged. That way, however, has itself been part of an approach to Christianity which we have met before. It is of a piece with the cast of mind which, for example, has wanted to recognize the Bible as divine revelation by claiming that the knowledge it contains can only be of divine origin. And as there have been those who have dismissed its divine message when they have found that its contents were available from human sources, so there have been those who have downgraded the sacraments when their direct institution by Christ has been in question. We should realize by now that this attitude is inadequate. We need to search out the influence of Christ in the establishing of these sacramental acts, acts whose significance the Church came to recognize and treasure. For we must always bear in mind that these sacraments which Christ has instituted, directly or indirectly, make us members of his body, that is to say, of the Church. Christ's sacraments are the sacraments of the Church. They bring us to a share in his life, death, and resurrection and so they bring to birth and maturity the community of those who believe in Christ. What baptism, confirmation, and eucharist initiate, penance purifies, and marriage and holy orders define more specifically; while the ministry of salvation is expressed in the sacrament of the sick. Let us look at each one in turn and ask about its origins, its purpose, and its effect.

(iii)

Baptism

The scriptural origins of baptism are plain enough. The custom of purifying rite is familiar to various religious traditions. Jesus was himself baptized by John before his public ministry began. The end of the Matthaean Gospel, as we have seen, presents him instructing his disciples to baptize. The Acts of the Apostles show that instruction being carried out. Pauline letters explain its significance. The biblical basis is strong.

Nowadays some parents are reluctant to have their children baptized. Remembering perhaps pressures which were brought to bear on them when they were young and conscious that the kind of deep commitment which the Church requires of its members should come from a free choice, they argue that baptism should be delayed until the individual can decide for himself or herself. It is a line of argument which shows up clearly the misunderstanding which has plagued our appreciation of baptism.

Through baptism we become members of the Church. The attitude I have just outlined sees the Church implicitly as a club with rules and regulations which constrain its members. But the Church should not be understood in those terms. It is a community of people, bound together by a common faith which makes them sharers in the life of Christ. To share in that life is to share in love. Christians belong to a community of love, and while all communities will have their rules for the sake of good order, rules do not have priority. Families need rules, but do not give them pride of place. Love comes first. How strange it would be were parents to say to their children, 'All right, we'll feed you and clothe you and put a roof over your heads, but we won't love you. However, if when you are sixteen, seventeen, or eighteen you decide you want to love us, then we will gladly love you in return.' We do not regard parents' love as an undue pressure likely to handicap a child's growth, nor even as an optional extra. On the contrary, we look sadly on those who have been orphaned because we know they have been deprived of something immensely valuable for their development as whole human beings. By baptism we are ushered into the known realm of God's love. If it has been experienced as a source of pressure—which has too often happened—then that is a tragedy. God loves all that he has made. He loves everyone, Christian and non-Christian, believer and unbeliever. We cannot escape his love. We may reject it, but he will never cease to love us. However, those who have been baptized have been privileged to have that love made manifest to them in particular words and through particular signs. In our discussion of the Church we have seen that the privilege entails responsibilities.

Besides the actual words of baptism in the name of the Father, Son, and Spirit, the baptismal ceremony contains various symbols. Oil, a lighted candle, and a white garment are all used to signify the life in Christ of the newly baptized. But the principal symbol is water. Water has many purposes in our lives and these create the associations which it symbolizes. We wash in water, and by our baptismal washing we are cleansed of our sins. We quench our thirst with water, and at baptism our thirst for God is being quenched. We pour water on to the earth to make plants grow, and this baptismal water is making us grow into Christ. Most fundamentally of all, however, by our passage through water there is symbolized our share in the death and resurrection of the Christ. That is expressed most vividly, of course, when baptism is administered by immersion. The plunging into the water is a sign of death and burial; the emergence a sign of resurrection. Even when water is only poured over the heads of those being baptized, it should remind us of the Israelites passing through the waters of the Red Sea from slavery in Egypt to freedom in the promised land. Their exodus anticipates our deliverance in Christ.

Living the Christian life means far more than learning about Jesus of Nazareth, admiring him, and following his example. We have spoken of sharing in Christ's life. What does that sharing imply? It cannot be reduced to the imitation of one whom we respect. That denotes only an external relationship. Sharing his life signifies an interior bond. That bond, we believe, is established through baptism. The symbol makes real what it represents. St Paul explained to the Romans what happens in baptism. 'Do you not know', he writes to them, 'that all of us who have been baptized into Christ Jesus were baptized into his death? We were buried therefore with him by baptism into death, so that as Christ was raised from the dead by the glory of the Father, we too might walk in newness of life' (Rom. 6: 3–4). Baptism lies at the basis of Christian life. It symbolizes our birth into a share in Christ's life. Its representation of his death and resurrection is efficacious. It makes real what it symbolizes and it makes them real within us. That death and resurrection are stamped upon us like a seal and establish our identity as Christians.

Confirmation

The sacraments of baptism and confirmation are closely related. In fact some people have argued that they are not two distinct sacraments at all. At confirmation it is said that we receive the Holy Spirit, but as we receive the Spirit at our baptism, there are those who conclude that there can be nothing more to be bestowed. If you have the Spirit, you have the Spirit. There is nothing further to be said. On the other hand, we can observe that Christian living is not something that is established within us instantaneously, either as individuals or as the community. We have to grow. We observe how people come to maturity in the Spirit. Even good people develop and become better. It is God's power at work in them. And we recall how in the Acts of the Apostles Peter and John go off to Samaria where the gospel has been preached and well received. We are told that these two prayed for the new converts 'that they might receive the Holy Spirit; for it had not yet fallen on any of them, but they had only been baptized in the name of the Lord Jesus'. Baptism had been administered, but this further experience of the Spirit had not yet occurred. 'Then', we are told, Peter and John 'laid their hands on them and they received the Holy Spirit' (Acts 8: 14–17). Scripture and experience indicate something distinct from baptism, although intimately associated with it. That is what we should expect. The person, newly born and mature, is the same person, but at distinct stages of life. Birth and maturity are not to be identified, but neither are they altogether distinct from each other.

The understanding of confirmation as the sacrament of Christian maturity has gained much currency recently. It is easy to see why. It is associated traditionally with the indwelling of the Holy Spirit in a particularly complete manner. This way of speaking, however, can be both a help and a handicap. As a help, it draws out our experience in a telling manner. In this context, for example, I think of my attitude to one of my sisters. Slightly older than me, she has, so far as I am concerned, always been a part of my life. When I was a boy, if anyone had asked me whether I loved her, I would have answered, 'Of course.' It never occurred to me that a brother might not love his sister. But I remember very distinctly, when I was about fifteen or sixteen, suddenly realizing one day that I really did love her and, if she had not been my sister, I hoped she would still have been my friend. Now confirmation corresponds in part to that experience in religious matters: what is familiar and has been taken for granted is suddenly perceived with fresh eyes and affirmed deliberately and consciously. Our faith needs to come alive in that way and it is not entirely mistaken to look upon confirmation as the sacrament which signals that awakening.

However, this approach can become a handicap if adopted too exclusively, for it seems to understand what is taking place entirely in human terms. But the sacraments are God's way of treating us, his way of loving us. When we are confirmed, he is giving us his Spirit whose gifts bring us to fullness of life in him. Confirmation completes what baptism has established, and it does so for a purpose.

Those who receive it have the status of witnesses and so they are, as the Second Vatican Council stressed, 'more strictly obliged to spread the faith by word and deed' (*Lumen Gentium*, 11). It is a sacrament of mission within the Church.

Eucharist

We have referred already to the origins of the eucharist at the last supper, and in the next chapter we will be looking at this sacrament in more detail. Our comments here can be fairly brief.

This sacrament and the sacrament of penance are distinguished from the other five by the frequency with which they are celebrated. In the eucharist Catholics believe that Christ is really present. The manner of that presence has been a source of acrimonious debate in the past, but the search for Christian unity more recently has changed the temper of the controversy. A more positive, a warmer —indeed, a more Christian—attitude has prevailed and uncovered a far larger area of agreement than previous generations had supposed possible. Some, of course, still resist the idea of the real presence, but very few actually believe in the real absence. Christians generally recognize Christ's presence to them in this sacrament.

It is a sacrament of food and drink. The bread and wine through which Christ is present symbolize the feeding of the Church. Christians are nourished by him and so their share in his life is built up. But we should notice something further. This rite was known to the early Christians as the breaking of bread. The loaf was broken and distributed; the wine was poured out and shared. That breaking and pouring out symbolize a body broken and blood shed in sacrifice. The Christ gave himself up for us, sacrificed himself to save us from our sins. That sacrifice is made present to us in this sacrament. By eating and drinking we are not merely sharing in his life; we are moreover sharing in a life in which he died and rose again for our sakes. The eucharist makes us sharers in the life of the crucified and risen Christ. What has been established in baptism and sealed through confirmation is nourished through the sustaining power of the eucharist.

The sacrament of penance

We have noticed earlier some of the questions which have arisen concerning the origins of this sacrament. It is a sacrament of which the first Christians appear unaware. How, therefore, can it enjoy the status of a sacrament? We need to search out Christ's influence upon its origins.

After Pentecost the first Christians believed that Christ would return and bring about the end of time, the consummation of all things, very soon. One of the most acute crises with which the young community had to deal was the dawning realization that this expectation was not about to be fulfilled. Some people had given up work in the belief that the Lord would soon be with them in his glory. Paul had to issue rebukes and warnings (see 2 Thess. 3: 6–13). Paul's own teaching on marriage and the single state is affected by this expectation (see 1 Cor. 7: 25–31). There were worries about those who had died before the Christ's second coming (see 1 Thess. 4: 13–18). And there was another consequence of the Lord's delayed return.

The first Christians had supposed that the gospel must be proclaimed urgently to as many as possible before Jesus came again. Those who believed and were converted were then baptized. Their sins were blotted out. Once the Lord returned in glory, they would receive their salvation. But time passed and converts were not always faithful. Baptism was not a recipe for perfection. People sinned again, sometimes grievously. They sinned again, but then they repented. What was to be the destiny of such people? Jesus was their saviour. He wished them to be reconciled to the Father. Had their backsliding frustrated completely his sacrifice for them?

The Gospels suggest a different conclusion. By turning to the Gospels the Church came to recognize that there must be a way for Christians who had sinned to be restored. The ministry of Jesus is alive with the concern for the forgiveness of sins. Immediately in view, of course, is the forgiveness bestowed through baptism, but Jesus often shows his sensitive awareness of human frailty. At one point he instructs Peter that he is to forgive the one who has wronged him not seven times, but seventy times seven. He does not mean four hundred and ninety times. It is a way of saying that he must always show forgiveness. God does not set us higher standards than his own. The baptized person who sins again cannot be beyond redemption. To suggest that he is does shocking violence to the whole drift and spirit of Jesus' teaching and mission. Provided he repents, the sinner is forgiven and reconciled to the Father once more. Like the eucharist, this sacrament sustains the Christian's share in the life of Christ; in particular, it restores him when he has fallen away.

Here too God's keen love is evident. For we do not imagine that God is forgiving us only when we celebrate this sacrament. Not at all. God is forgiving us constantly. But reflection on our own experience can be instructive again. When we offend those who love us, their love instinctively exercises forgiveness. And we recognize that. Knowing that we are forgiven, however, does not remove our desire to express our sorrow. On the contrary, we wish to show that we are sorry all the more. The sacrament of penance gives us the opportunity to experience God's forgiveness in a way that is considerate of our condition. It gives it a particular setting and it also permits us to express our sorrow. This interplay between forgiveness and repentance can also be creative.

I heard some years ago of a German Jewish couple who escaped from Germany to America when the Nazi campaign against the Jews intensified. Their families were not so fortunate and died in concentration camps. Some time after the war the husband met a fellow German one day. This man was newly arrived in America and so the husband invited him to dinner that night. As it happened, the wife was unwell and so the two men ate together by themselves. Late in the evening, conversation turned to the war. It emerged that the guest had been an officer in various concentration camps, exterminating Jews. Unaware that his host was a Jew, he talked on and, when pressed about his attitude to what he had done, replied simply and calmly that he was an army officer obeying orders. As far as he was concerned, that was the end of the matter. He had no sympathy with those who tried to make some immense moral issue out of it all. He felt no guilt or regret for what he had done. The talk went on, further details emerged, the camps where he had worked and the dates. Then the host excused himself for a minute or two and went upstairs to see his wife. He managed to persuade her to come down and when she entered the room he said to her, 'Here is the man who was responsible for the deaths of your father and mother and brothers.' And the woman went up to her guest. She put her arms round him and said, 'You poor man.' And he broke down and wept. Only then, when he was forgiven, could he recognize that he had done wrong. The sacrament of penance is so like that. It should be administered as the sign of God's forgiveness. To experience such forgiveness is to discover deeper self-knowledge and can move us to repentance. The sacraments are creative. They bring about what they symbolize. It is sad that recently fewer people have been making use of this sacrament. We know that we sin. When we express our sorrow and experience forgiveness, we discover God at work, healing our sins and reconciling us to himself.

Marriage

The origins of marriage as a sacrament are complex. It was not regarded as one until the ninth century. Those passages in scripture, like the miracle at the wedding feast at Cana, which have so often been used as evidence of a biblical basis and institution by Christ, prove, as we have seen, to be misapplied. They are concerned with something else. Once again, however, we need to be on the alert. We should ask ourselves what it is we are looking for. Ours is not a blueprint Christianity. Our faith was not handed to us in a pure, idealized form. How did Christians come to give marriage its exalted status?

The question could receive a lengthy, historical answer. That is not our purpose here. But we can look at the Bible and discover there the prominence which this relationship receives. In both accounts of human creation, man is made whole through the making of male and female (Gen. 1: 27; 2: 20–4). The Decalogue forbids adultery and the coveting of a neighbour's wife (Exod. 19: 14, 17). Prophetic literature uses marriage in various powerful passages to bring home to the people the nature of their relationship with God. We may remember most easily the prophecy of Hosea (1: 2–3: 3) and the allegorical history of Israel in the Book of Ezekial (16: 1–63). These are obvious random examples.

Jesus too is shown giving great prominence to the married state in his teaching. When questioned about divorce, he reaffirmed the unity of man and woman in marriage. This is the natural, primary human state. He acknowledged later in private to his disciples that there might be reasons—sometimes most noble ones—why people did not marry, but the drift of the passage is plain. Generally speaking, marriage has primacy (Matt. 19: 3–12). And in the letter to the Ephesians the bond between husband and wife is described as illuminating the bond between Christ and the Church (Eph. 5: 21–33). The passage, of course, has had a harsh press recently. Wives are told to be subject to their husbands. It is easy to overlook the command that husbands and wives are told that they should obey one another and that, more specifically, husbands should love their wives, *as Christ loved the Church*. Christ loved the Church by giving up his life for us. These passages and others like them illustrate the influence teaching on marriage has had on Christian living. And there is something further.

At the heart of Christian life is the command to love. No one is beyond the boundaries of this loving. It is inclusive. But we are all suspicious—and rightly so—of the vague philanthropist, the person who loves everybody and so in fact loves nobody. If our love is to be real, then it must be earthed in the particular. In marriage supremely this great commandment of love can be fulfilled. There is no conflict in fact between the inclusive love of Christ's command and the exclusive love of Christ's sacrament. The sacrament enables the command to be realized.

It is not surprising that in the light of such reflections the Church should have come to look at marriage in a fresh way and to perceive not just a particular kind of life, but a reality so profoundly taken up in loving that it was a sign, a symbol, a sacrament of the love of Christ for the Church. In and through marriage we can glimpse and experience Christ's love for us. Non-Christians may sometimes ask what all this business of Christ's love for the Church is about. It can seem such an abstract notion. We should be able to point to committed married Christians. 'Do you see their love for each other,' we should be able to say, 'their understanding, their patience, their kindness, their readiness to forgive? Then you catch a reflection of the love between Christ and the Church.'

One final thought on marriage. Here too people share in the death and resurrection of Jesus. Sometimes one person will love another so deeply that the possibility of ever having married anyone else is completely excluded. More usually, though, however happy we may be, we will recognize amongst our friends or acquaintances someone to whom, we suspect, we could have been married very contentedly. If our marriage is going through a drab phase, the other person may even appear to make us alive as no one else whom we have ever met has done. That is generally an illusion, of course, not least because in any fresh relationship we would always have a part; as a friend of mine once remarked, 'I used to wonder what it would have been like to have married someone else, I thought how different it would be. Then I realized, no, it wouldn't, because there would still have been me.' But such awareness can bring home to us that we choose a particular path in marriage and exclude all others. There is a sacrifice in that, a real dying to self. And we have to die to ourselves as well, even in the most unclouded loves, because we must overcome the self-interest which is rampant in us all in order to make real the unity and wholeness which married love should proclaim. The surrender of selfish whims and ways will hurt. How humbling to be so put out by pinpricks. Accepted with the generous and open heart, however, the risen life floods in.

Holy Orders

If we look for the origins of the ordained ministry in some neat biblical text, we look in vain. Once more the beginning was not like that. But the Gospels relate perfectly plainly that Jesus called certain individuals to fulfil a particular role in helping him in his public ministry. Disciples gathered round him and from these he summoned twelve to be his apostles. There is a solemnity about their calling. Matthew and Mark speak of them having authority—a powerful word—over unclean spirits to cast them out, and of their duties of preaching and healing. In Matthew they are called just before being sent out on their mission (Matt. 10); in Mark we are told that he went up into the hills 'and called to him those whom he desired' (Mark 3: 13). The Lucan text does not speak of authority and duties at this point. Instead it emphasizes the time Jesus spent in prayer and how 'all night he continued in prayer to God'; and at once it continues: 'And when it was day, he called his disciples, and chose from them twelve, whom he named apostles' (Luke 6: 12–13).

We learn from the Acts of the Apostles that later daily practical matters were diverting the twelve from their principal responsibilities, and so they saw the need for another kind of service. They appointed seven men to these tasks. They called them servants. We have anglicized the Greek word and know them as deacons. They were not appointed casually or merely out of convenience. A need had been recognized. The community set them before the apostles 'and they prayed and laid their hands upon them' (Acts 6: 1–6). The note of solemnity is unmistakable.

Then in the early stages of the lives of these first Christian communities there were elders. Their status has been the subject of dispute. Were they really presbyters or bishops or simply local people doing a job without any special commissioning at all? It is difficult to give a cast-iron answer, but the Pastoral Epistles to Timothy and Titus give us evidence of an emerging church order which by the time Ignatius was writing, around 117, had at least in certain areas become defined quite clearly. The bishop was the leader of the local church, with presbyters and deacons gathered round him. Once again, the human elements which can be traced in the development of this church order are no necessary obstacle to its place in the divine plan. As with the Petrine ministry, need and circumstances would be crucial influences upon the recognition of its status.

The clergy are not 'the best Christians'. The title of 'other Christ', so often used to describe the priest in the past, is not favoured today because it obscures the great truth that every baptized person is another Christ. Indeed all share in the ministry of the one Christ, the great high priest. The difficulty for us is the use of the word 'priest' to translate different Latin words, *sacerdos* and *presbyter*. Christ as the great high priest is the *Sacerdos Magnus*. The word means the one who gives—and so makes—sacrifice. We have seen that every Christian through baptism is drawn into the death and resurrection of Christ; every Christian makes that sacrifice. All are *sacerdotes*, sharing in the priesthood of the great *sacerdos*. The presbyters, however, are another matter. By virtue of their baptism they, like every other Christian, are *sacerdotes*; but they are presbyters by virtue of their ordination. As presbyters they are meant to give a pastoral leadership. They have the task of serving the community through their ministry of the Word and celebration of sacraments: word and sacrament are aspects of the same single reality. In this way they try to bring harmony to the Church. Or, to put it differently, they are the ones who try to bring to the Church holy order. You will realize at once the implications of that work. A community which is ordered properly, in the way appropriate to itself, is being true to itself. The presbyters' ministry, therefore, assists the self-knowledge of the Church. Another familiar theme recurs.

This ministry is demanding work. The Gospel contains many warnings that the servant is not greater than his master and that the disciple must in his turn take up his cross and follow Christ. But it is also rewarding. The faithful disciples are told, not only that they will inherit eternal life, but that they will receive far more than they have given up 'now in this time' (Mark 10: 30; Luke 18: 30). Here again the pattern of death and resurrection is inescapable.

The Anointing of the Sick

This sacrament has been the Cinderella of the sacraments, hidden away, almost unmentionable. I do not wish to exaggerate, but it is true that in the past it was associated almost exclusively with death, so that the arrival of the priest was often understood as the sign that the patient's condition was hopeless. The emphasis has been changed. It is celebrated now when a person is seriously ill (so not just for coughs and influenza), but not only when their condition is grave. It has been regarded formally as a sacrament only since the twelfth century, but the practice of anointing goes back to the earliest years of the Christian community. Its purpose was partly medical and partly spiritual. I think of it as the sacrament of Christ's saving work.

When we consider our history, our original estrangement from God, the calling of the chosen people, the ebb and flow of their relationship with God through patriarchs, Kings, and prophets, and the climax of this drama with the coming of the Christ, we see a work of reconciliation, of salvation. The Lord came that our sins might be forgiven. It is natural to say that he came to heal us. And so we are not surprised to find that the forgiveness that he proclaims throughout the Gospels is accompanied by miracles of healing. The account of the man lowered from the roof in front of Jesus is typical. He forgives the paralytic his sins. Scribes who are present regard these words as blasphemous. Jesus challenges them: 'Which is easier, to say to the paralytic, "Your sins are forgiven," or to say, "Rise, take up your pallet and walk"?' Without waiting for a reply he continues, 'But that you may know that the Son of man has authority on earth to forgive sins', and now he addresses the sick man, 'I say to you, rise, take up your pallet and go home.' And the paralytic obeys at once (Mark 2: 1–12). There are many other examples. Then a little later the disciples are sent out and we are told that they 'anointed with oil many that were sick and healed them' (Mark 6: 13). What Jesus had done, they also were to do. And finally we may remember that passage in the Letter of James: 'Is any among you sick? Let him call for the elders of the Church, and let them pray over him, anointing him with oil in the name of the Lord; and the prayer of faith will save the sick man, and the Lord will raise him up; and if he has committed sins, he will be forgiven' (Jas. 5: 14–15). What the disciples had done was becoming the established practice of the early Church. And in the fifth century Pope Leo the Great said as much: 'What was manifest in Christ, now takes sacramental form in the Church.'

Sacraments are not magic. This sacrament does not make healing automatic. Every sacrament is an act of God's love for us. Illness is ambiguous and sooner or later death is unavoidable. God cares for us unceasingly. In this sacrament his care for us when we are sick finds a particular expression. Through this sacrament we are prepared for death or recovery, whichever the case may be. By this sacrament also we are drawn into the death and resurrection of the Christ: the care for the sick is an outward sign of the care for those who are sick in spirit; their care was his saving work.

Conclusion

To be a Christian is not to be a mere imitator of Christ. Jesus is not a hero to be set on a pedestal for admiration from afar. He is our friend, our brother, and we are called to share in his life individually and as a community. That sharing is not a fanciful notion, empty of content. We are bound to him in love and his love is displayed for us and has its effect on us through these sacramental acts. They draw us into a deeper share in his life, death, and resurrection. They make us Christian, they make us Church. They are acts of God's love for us and they are celebrated in the Church and by the Church and so they bring the Church to life. By baptism we are reborn and established in the new life of the believing community which his death and resurrection have won for us. In confirmation we are commissioned to share this life with others. The eucharist nourishes it; the sacrament of penance purifies it and builds us up in it. Through the vocations to marriage and holy orders we bear witness to that life in us and for others. And in the sacrament of the sick we show forth that care and love which are fundamental to the saviour's ministry.

And so, finally, a trick question: in which sacrament is Christ really present? Catholics will answer, the eucharist. But the question is a trick for in fact Christ is really present in every sacrament. It is true that he is not identified with the water in baptism, for example, or the oil in confirmation, but the reality of his presence in these acts should never be in doubt. However, to be fair, there is, of course, a sense in which Catholics believe that Christ's presence in the eucharist is of a special kind. We can now consider that belief and, furthermore, the celebration of the mass which lies at the heart of our worship.

9

Celebrating the mass

(i)

ONE of my most vivid memories from my days as a student in Rome came from my attending some of the public sessions of the Second Vatican Council. The Council met in private, but on the days when the documents it had composed were to be promulgated, there were these public sessions. I went along when I could. The Pope, Paul VI, would concelebrate mass with a dozen or more bishops. The other bishops were in attendance and the rest of the basilica was crowded with humanity of many sorts and shapes and sizes. And as the Pope began the mass and this vast congregation thundered their responses, it always seemed to me that here in a quite exceptional way the Church was at prayer. It brought home to me the saying: the Church is never more truly itself than when the mass is being celebrated. In St Peter's during an ecumenical council that truth is evoked most powerfully but it is equally true at every celebration of the mass. We can seek out the reason.

The existence of the Christian Church is absolutely dependent upon Jesus Christ. That statement may seem blindingly obvious, but we have to realize what it implies. This community came into being through its faith in Jesus. He was the one, at once human and divine, who was perfectly faithful to the will of his Father in heaven and who in fidelity to that will went to his death on the cross and through his fidelity was raised from the dead. We believe in those events. Our existence as a community of faith rests upon them. His life and death and resurrection make us Christian, give us our identity. They are constitutive of us as the Christian people. If he had not lived and died and risen from the dead, our religion would be an illusion. We depend upon his sacrifice.

We know what constitutive actions are. Although a debating society, for example, may hold dinners and arrange outings, perhaps annually to the House of Commons, it is most truly itself when debating. That is its constitutive action. Debating makes the society what it is. The mass is that and more than that for Christians. It not only celebrates what we are, but it does so by taking us into the living presence of those events which give us our identity. When a mass is celebrated, the life, death, and resurrection of Jesus are being remembered and made present. They are not being remembered in a feeble, passive way; nor are we gazing back nostalgically at a past act from the earliest days of our history. This memorial is active. We return to our origins. It makes present the very reality which constitutes us as Christian.

What we may find difficult about this teaching as we first approach it, is understanding how it does not involve something repetitious. It may seem that we are going over the events of Christ's passion again and again, as though something had been lacking before. But Christ, we know, died on the cross 'once and for all'. One part of the solution to this problem has been mentioned already. We have to rid ourselves of the assumption that memorials are only passive and realize that through a memorial which is active we enter into what is remembered, and so, in this instance, come to share in the redeeming work of Christ; the effects of that work become our own. And another part of the solution takes us back to a point we considered much earlier.

When we were thinking about the passion of Jesus you may remember we stressed that the cross and resurrection were not to be regarded on their own in isolation, as though they were the acts which won our salvation, independently of anything else. We saw them rather as the climactic outward manifestation of that interior disposition of utter fidelity which dominated everything Jesus did. That fidelity is absolutely fundamental. And at the same time the cross and resurrection are not to be separated from it. They are its absorbing expression; they are woven into it inextricably. They are its consequence. As we share in Christ's life through our own fidelity, so will we die and rise again.

The mass which is this active memorial of our redemption recalls what took place at the last supper: 'Do this in memory of me.' It is vital for us because of the significance with which that supper was invested: it proclaims and celebrates at the same time the fidelity of Jesus which was revealed at its peak in his dying and rising. By participating in the mass we seek to deepen our own fidelity. It draws us into his death and resurrection. In the mass we are plunged into the midst of a present reality. The living and the dying and the rising which alone make us Christian become present to us in a sacramental way. We share in them. By doing so we take part in the reality which makes us what we are. It is a perpetual reality. Our actions are not repetitious. Through the mass we are able to return to the living source of our existence. It is the great act of our identity. That is why the Church is most perfectly itself when a mass is celebrated.

Let us look a little more closely at what takes place.

(ii)

The mass begins with a brief introduction and rite of penance; it concludes with a blessing and dismissal. Otherwise it is divided into two main parts and each part has two subsections. The first part revolves around the reading of scripture and is called the liturgy of the word; the second focuses on the eucharist and is known, therefore, as the eucharistic liturgy. These parts run parallel to each other. Each contains a consecration and a communion.

At first, then, passages from the scriptures are read out. Consider what happens when readings—any readings—occur. When a book is closed or left open but neglected, the words it contains are no more than dead marks on the page. However, once a person begins to read those words and especially when he begins to read them aloud in the presence of others, they come alive. They touch those who listen and the meaning they communicate comes to be present amongst those who hear and understand. That much is plain enough.

Now we need to remember also that when the scriptures are being read their message is for Christians the word of God. Moreover, as we realized earlier (p. 58 above), these words are not just words, not mere propositions; these words are the Word. The message is a person. The person is Christ. When the Bible is read out, the words come alive and they speak of him. The Christ is made present. Here in the liturgy of the word we find the consecration. For this reason these readings should be carried out with the greatest care. We would be scandalized by clergy who consecrated the sacred elements in a slovenly manner. We expect of them the highest standards for their duties. We should have the same expectations for the way in which the scriptures are read during mass.

A homily normally follows. These words which the preacher prepares are not meant to be a harping on any chance subject. They are meant to unfold for those who are present the meaning—or at least one aspect of the meaning, however slight—of the reading which has just taken place. To use the natural word, the preacher has to *communicate* that meaning to the people. They are to have communion with the Christ who has been made present to them by the reading of the Word. Consecration and communion. The one who became present through the proclamation of scripture dwells in the hearts of the faithful through the words which are preached.

We pass on to the eucharistic liturgy. Here again there is a consecration and a communion. In the eucharistic prayer the priest speaks the narrative of consecration. He speaks of Jesus on the night before he died, taking bread, blessing it, breaking it, and giving it to his disciples, saying, 'Take, eat, this is my body'; and in the same way, taking the cup, giving thanks, and giving it to his disciples, saying, 'Take, drink, this is my blood.' These actions described are not like stage directions. You sometimes see clergy cracking the host at the reference to the breaking. That is a mistake. There is an appropriate time for the fulfilling of each of these actions during the mass. They are not stage directions, but the narrative itself is dynamic. What does that mean? It means that the words are powerful words and they bring into effect what they state. They are not a magical formula, like 'abracadabra', a sound which signals a trick. Only God's words have this power. Our ancestors in the faith understood this truth and we find it set forth most dramatically in the first account of creation: 'God *said*, "Let there be light . . . a firmament . . . dry land . . . "', and so on. What he spoke came into existence by his speaking. His word was powerful. What it stated came to be. And when the priest utters the narrative of institution, because he is the minister of God he is not merely making intelligible sounds, but the words he speaks, because he is ordained so to speak, are dynamic: they bring into effect what they signify. Bread and wine are consecrated and Christ becomes really present among his people.

What has been taken and blessed is shortly afterwards broken and given. The Christ, sacramentally present, is communicated to the people. The consecration of the elements leads on to their distribution at communion.

Thus both sections of the mass run parallel. In both there is a consecration by which in different ways the Christ becomes present and in both, once present, he is communicated to the people.

Sometimes Catholics express surprise after attending the eucharist of another Christian tradition. 'It's so like ours,' they say. Well, so it ought to be. Every authentic Christian eucharist should reflect this pattern. And its origins, of course, are scriptural. I am not referring only to the last supper, but to the way in which our worship developed.

Think of the account of the two disciples on the road to Emmaus. They are joined by Jesus whom they fail to recognize. They talk. He questions them and draws from them a statement of the events of the previous week, their dashed hopes, and their perplexity at the women's announcement that he is risen. What they are saying to him, in other words, is like what we listen to when the scriptures are being read out. Jesus then says to them: 'O foolish men, and slow of heart to believe all that the prophets have spoken! Was it not necessary that the Christ should suffer these things and enter into his glory?' And he goes on to unfold for them the significance of what they have just been saying to him. In effect, he preaches them a homily.

They reach their destination. He seems to be going further. They prevail upon him to stay with them. The next verse has a solemnity about it. We can be in no doubt as to what is taking place: 'When he was at table with them, he took the bread and blessed and broke it, and gave it to them.' The four actions are specified. Eucharistic consecration and communion have taken place. So here in this Lucan text (Luke 24: 13–31) we may say that Jesus is described as celebrating a kind of extended mass with these two disciples. All the essential ingredients are mentioned except, we might add, the cup.

There is one further point. Immediately after Jesus vanishes from their sight, we are told that they say to each other, 'Did not our hearts burn within us while he talked to us on the road, while he opened to us the scriptures?' (Luke 24: 32). If their hearts burnt within them, then they burnt within them, the point being that they became aware of the burning only after the breaking of bread. It indicates the wholeness of what has happened. It does not mean that after receiving communion we should remember a fiery sensation during the readings, but it should encourage us to participate as fully as possible in the mass so as to share more deeply in the reality which is taking place. Word and sacrament are profoundly at one.

(iii)

Let us consider next the belief that the Christ is really present in the eucharist. If it is true that very few Christians believe in his real absence from that sacrament, nevertheless it is a teaching which has given rise to deep and lasting divisions. What account of this belief can a Catholic give? Long familiarity may have made its affirmation easy, but there are indeed times when I, as a priest, look down at the altar during mass and am almost overwhelmed by my belief that the bread and wine which I have just consecrated is bread and wine no longer, but the body and blood of Christ the Lord now really present. What can be said?

It is helpful to remember the exodus. The Bible tells us about the enslavement of the Israelites in Egypt and their deliverance. We do not need to delve here into the details of the actual historical events. We can remain with the biblical account. There we are told that when the night of their liberation came they were to eat a meal, no doubt in part to be nourished for the journey. They were to eat a lamb without blemish and mark the lintel and doorposts with its blood. That would protect them from the angel of death who was going to destroy the first-born of the Egyptians. They were to eat unleavened bread as well. There was a cup of blessing also. These instructions were fulfilled. The people were delivered. They passed over from their physical slavery in Egypt to the freedom of the promised land. There each year they celebrated their deliverance by eating again that passover meal as a memorial. It was not passive remembering. The memorial took them back to their roots; it affirmed their identity as the people God had chosen as his own.

We come forward many centuries. In an upper room in Jerusalem Jesus is celebrating that passover meal with his disciples. (Some people have questioned whether the last supper was in fact the passover meal. The point need not detain us. At the very least, the supper was soon seen in that light.) And at the appropriate moments in the meal, when the unleavened bread is broken and distributed and the cup passed round, we are told that Jesus paused and identified himself with these elements: 'This is my body which is for you . . . This cup is the new covenant in my blood . . . ' (1 Cor. 11: 24–5). It is impossible to imagine the bewilderment which must have come over the apostles. To them his words must have seemed meaningless.

After the supper they go out. We know what followed. Jesus was arrested, tortured, and executed. Then on the third day he rose from the dead and they saw him. Gradually the significance of these events dawned on them. Their ancestors had been saved from slavery in Egypt; they had passed over from that slavery to freedom. The risen Christ revealed the salvation of which that earlier deliverance had been only the anticipation. Death had been defeated. And the significance of the supper dawned on them as well. The meal which had celebrated their ancestors' passover had been transformed. This new passing over from sin and death to new life in the risen Lord had been expressed in this action: Christ was the lamb who was slain and whose blood daubed, not lintel and doorposts, but the wood of the cross, so that all people might be delivered from their slavery to sin and pass over in freedom to the promised land and life of salvation. The unleavened bread and the cup of blessing find their meaning fulfilled by their being identified as his body and blood.

It is essential to remember this background if we are to glimpse what our belief in the eucharist proclaims. When its origins in the history of the Jews are ignored, it can seem so easily to be arbitrary and unbelievable. The origins in no way explain the mystery of the belief, but they put it in a context which allows us to see more readily what is taking place. The mechanics of a mystery are always elusive. We may nevertheless recognize its truth.

In this case we find St Paul writing to the Corinthians about the last supper no more than thirty years or so after the crucifixion. It is fascinating that the style and tone of the passage make it evident that he is not suddenly telling them about some fresh teaching which has just occurred to him. He is reminding them of something which he had himself received and which they already knew: 'For I received from the Lord what I also delivered to you . . . ' He goes on and gives his account of the last supper: 'The Lord Jesus on the night when he was betrayed took bread, and when he had given thanks, he broke it, and said, "This is my body which is for you. Do this in remembrance of me." In the same way also the cup, after supper, saying, "This cup is the new covenant in my blood. Do this, as often as you drink it, in remembrance of me."' And he concludes 'Whoever, therefore, eats the bread or drinks the cup of the Lord in an unworthy manner will be guilty of profaning the body and blood of the Lord' (see 1 Cor. 11: 23–7). No one, of course, should try to force this passage to mean more than it says. We do not need to

uppose that Paul could have written a medieval treatise on transub-
:antiation. At the same time, the direct association of the bread and
vine as the body and blood of Christ is inescapable. By the middle of
ae following century we find St Justin Martyr describing the
elebration of the eucharist and writing: 'For we do not receive this
ood as ordinary bread and ordinary drink; . . . we are taught that
ae food over which the prayer of thanksgiving, the word received
om Christ, has been said, the food which nourishes our flesh and
lood by assimilation, is the flesh and blood of this Jesus who became
esh' (Justin Martyr, *The First Apology*, 66).

From such teaching the belief in Christ's real presence in the
ucharist emerged and grew strong. The term 'transubstantiation'
vas used by the Fourth Lateran Council in 1215 and later canonized
y the Council of Trent. In view of the controversy which has
urrounded it, it is good to grasp firmly what it was intended to
nsure. It has been put most succinctly at the end of a recent study:
The creed of Lateran IV was neither the culmination of twelfth-
entury eucharistic theology nor a prohibition against further specu-
ition about the mode of Christ's presence in the sacrament . . . It
as not the mode of presence which Innocent III wished to affirm
ere, but the presence itself' (Gary Macy, *The Theologies of the
'ucharist in the Early Scholastic Period* (Oxford, 1984), pp. 140–1). In
ther words, this term was not adopted in the Church because a
articular way of describing *how* Christ was present in the eucharist
as favoured; it was adopted simply as the most effective means that
ould be found for expressing *the fact* of that presence. We are back at
ae earlier point: we waste our time when we try to describe the
techanics of mystery; we should seek only to proclaim and safe-
uard the deep truth that it declares. We do not know how Christ is
resent in the eucharist; at the same time we believe in the fact of his
resence. We believe that he is really present, really present in a
acramental way. It may be useful to say something about that
acramental real presence.

(iv)

often seems that the reason many people find this teaching on
Christ's presence in the eucharist hard to accept derives from their
vay of thinking about it. They are handicapped by an approach
vhich is over-simple. They consider the idea of real presence and

decide that two people can only be really present to each other when
they are in each other's company. That is evidently a form of real
presence. But when my telephone rings and I speak to friends who
are calling me from Sydney, are they really present? Or when I look
at the fine eighteenth-century grandfather clock which stands in my
room in the Old Palace and which was given to the Oxford
Chaplaincy some years ago by Major Peter Middlemiss, am I
looking only at his gift? We sometimes call gifts presents, presumably because they make the giver present. Now I am not suggesting
that the presence of Christ in the eucharist is to be likened to a 'phone
call from Australia or a grandfather clock. I offer these two random
examples merely to help make the imagination more supple. There
are more ways of presence being real than the physical. One such
way is sacramental.

'Sacramental', too, however, can cause problems, especially for
Catholics. The sacramental is equated with the symbolic and the
symbolic with the token sign. In no time at all, a sacramental
presence is interpreted as completely external to the reality it
symbolizes. In the west, black is worn as a sign of mourning; in the
east, white. The choice of dress for the symbol is arbitrary. By now
readers of these pages should realize that there is another possibility.
The signs which give outward expression to the deep interior truths
and realities of our lives, like the kisses and declarations of love between lovers, are not simply detached and arbitrary. They are caught
up inextricably in what they express. They are part of the reality.
Such a sign is not external. It participates in the reality it conveys.

It helps to draw these two considerations closer together. When
we can realize that not every real presence is also physical and that
there are symbols which are not totally separate from the realities
they symbolize, then there may begin to dawn on us an idea of what
sacramental real presence means. Further depths can be investigated,
but at least we will be approaching the issue constructively.

Some people, of course, may feel threatened by this approach.
The teaching they have received from childhood, often accompanied
by stern warnings not to bite the host because they will be 'biting
Jesus', may well have made it particularly hard for them to accept the
sacramental character of the real presence in a way that does not
imply a diminution of the doctrine. But the teaching is not being
watered-down, and I think it is possible to illustrate this fact fairly
simply.

For many centuries until recently it has been the general custom in the Catholic Church for everyone except the celebrant to receive communion under one kind only at mass; the congregation did not receive communion from the chalice. But it was never believed that this manner of communicating meant reception of only a partial presence. We receive the Christ at communion, whole and entire. The host is not disguised flesh, the wine not camouflaged blood. The Catholic eucharist is not a form of cannibalism. It is a sacrament. The whole Christ is received, body and blood, soul and divinity, under either form. That can only be true if the manner of real presence is sacramental.

At the same time, it is worth adding that it is most desirable that everyone at mass should receive communion under both kinds, whenever possible. Rome has laid down certain occasions and conditions, but these must not be interpreted in a way contrary to the spirit of the liturgical renewal which has been taking place. To receive communion from the chalice as well as under the form of bread is to be drawn more deeply still into a share in the sacrifice and the supper of the Christ.

The community of the Church is most perfectly itself when mass is celebrated, but we know that we, the members of that community, are not yet perfect. We sin. We fall short. It is time to turn to the question of Christian morality.

10

Christian morality

(i)

ONE major theme of this account of Catholicism has been the link between the community's identity and its self-knowledge. Where self-knowledge has broken down, identity is lost. And a vital part of that self-knowledge will always be the moral sense, the ability to know what is right and what is wrong. For Catholics this moral sense is not to be understood simply in terms of doing good and avoiding evil. Very early in this account we considered the significance of Jesus for us as his drawing us into a share in his life. The birth of Jesus reveals to us the true meaning of being human. We attain that true humanity ourselves through our Christian discipleship: in other words, as followers of the Christ we are faithful to God. Fidelity is an interior reality which becomes manifest in our lives as we accept the consequences of obedience and love. We will be wounded deeply, but by remaining faithful we are healed. There is new joy. It is the pattern of death and resurrection. Those who are faithful are the Church and they are established and nurtured as the Church through the sacraments. To say so much is only to sum up from a particular vantage point everything that has been said in these pages so far. When we consider Christian morality, we are trying to uncover guidelines for Christian behaviour in the light of this life in Christ which we share. There is a warning in that as well. In one sense to describe this morality as Christian is misleading. It may suggest that we can outline a way of life that is exclusively and distinctively Christian, as though there is a pathway through life marked 'Christians Only'. But the birth of Christ and everything following from it has taught us a different lesson, one often repeated here: we are not to expect something exclusive; we must rather seek out the divine which is present in the human. So we are not looking for a way of action which only Christians follow, but that way of action which is most perfectly the outcome of lives rooted in Christ

(ii)

Before the Second Vatican Council Catholic teaching on morality was very preoccupied with sin. These were clearly categorized. Some were mortal: for example, to miss mass on a Sunday or Holyday of Obligation, to murder, to steal, to commit adultery, or to eat meat on a day of abstinence. These sins cut you off from God and, if you died without being sorry for them, you went to hell. Others were called venial: the neglect of daily prayers, lying, selfishness, bad temper, and so on. These did not sever the relationship with God, but their evil was not to be underestimated. Any offence against God is serious. It would be a travesty of earlier times to suggest that what was done was the only factor which was taken into consideration when assessing the morality of actions. Circumstances and intentions were considered, but their role was quite secondary. The actions themselves were the primary consideration. More recently, however, this approach has altered. People have become dissatisfied with a scheme of things which found it so hard to differentiate between the man who eats a steak on Good Friday and the man who is habitually unfaithful to his wife. It is not that these categories are useless, but their usefulness is limited.

One apparent advantage of such categories, of course, was the clear guidance which they seemed to give. Older Catholics sometimes look back on those days nostalgically. 'You knew where you were then' is their frequent refrain. At the same time, they berate the newer approach which, they say, has fudged the categories and made life easier.

Parish visiting in Wallasey some years ago, this view of affairs was set before me vigorously by a man in his sixties at whose house I had called. But there was one difference. Regretting the passing of these former ways, my host nevertheless reached another conclusion. 'The trouble is, Father,' he said to me, 'it makes life so much more difficult.' I was, and remain, impressed. It is so wearying to have the teaching of the Christian life which the Second Vatican Council expounded, constantly caricatured as the Church gone soft or making things too easy, when in fact a far more demanding vision was being presented. And problems arise as we adjust to it. The earlier way had a clarity resembling the approach which children need: they have to have a definite idea of right and wrong. But we recognize that adults have to think in less simplified terms, which are what the Council has tried to encourage. But as in life generally, so here. The transition from childlike simplicity to adult complexity is not achieved in an instant. The process is gradual. As people have to pass through adolescence to maturity, so does the Christian community as it strives for maturity in Christ. The problems of adjustment are characteristic of the ebb and flow of the condition. They should not be made into grounds for despondency, still less for a retreat to the simplicities of childhood once more. As we seek an ever fuller life in Christ, so we must face up to the demands such a life will make upon us.

(iii)

To live in Christ is to live a life of the purest, unqualified love. Love lies at the very heart of Christian life. Everyone should know that. When Jesus was asked which commandment was the first of all, we are told that he directed the enquirer's attention to the words of Deuteronomy: 'Hear, O Israel: The Lord our God, the Lord is one; and you shall love the Lord your God with all your heart, and with all your soul, and with all your mind, and with all your strength.' And he added a second commandment: 'You shall love your neighbour as yourself' (Mark 12: 29–31; see Deut. 6: 4). These words were echoed in the Gospels of Matthew (Matt. 22: 37–9) and Luke (Luke 10: 27), and were phrased afresh in the Gospel of John at the last supper when Jesus spoke to the apostles of a new commandment of love. They were to love one another as they had been loved by him, and he compared his love for them with the love he had himself received from his Father: 'As the Father has loved me, so have I loved you . . . This is my commandment, that you love one another as I have loved you' (John 15: 9, 12). The teaching is clear. Paul caught it up in his letter to the Romans when he told them, 'Owe no one anything, except to love one another; for he who loves his neighbour has fulfilled the law . . . love is the fulfilling of the law' (Rom. 12: 8, 10). And the first letter of John expressed pungently the unity of the two commandments: 'If anyone says, "I love God," and hates his brother, he is a liar; for he who does not love his brother whom he has seen, cannot love God whom he has not seen' (1 John 4: 20). So pervasive is this theme that we have touched on it explicitly already when reflecting upon sacramental life (p. 81 above) and we will return to it later (p. 134 below). Indeed, we have seen that Christ's fidelity is a fidelity of obedience and love; that the Church is called to just such fidelity; that the sacraments introduce us into love, establish us in love, and build us up in love; and when we consider the life of virtue, we remember that while faith, hope, and love abide, 'the greatest of these is love' (1 Cor. 13: 13). Love is the vital source of Christian living.

It is only when we allow the power of this commandment to love to break in on us that the horror of sin can be perceived truly. We are called to love. We are true to ourselves only as loving men and women. But sin attacks love and so it attacks the roots of our identity. Nowhere is this illustrated more graphically than in the Gospel accounts of Jesus' temptations in the wilderness. They are often spoken of as temptations for Jesus, so to speak, to slip his disguise, revealing his true self, turning a stone into bread or summoning the angels to safeguard his leap from the pinnacle of the temple. But this interpretation is mistaken. The third temptation, to have power over all the world's kingdoms if he will worship the devil, shows us what is happening most clearly. These are not temptations to reveal his true self; they are temptations to destroy that self. What kind of Messiah would he have been if he had resorted to a trick of magic, put the Father to the test, or been greedy for a power which was already his? To have yielded to the temptation would have been to cast off his messiahship. The same is true whenever we are tempted and sin. Our true identity, rooted in love, is being eroded.

Ponder carefully some of love's most outstanding qualities. Love is service. 'The Son of Man came not to be served but to serve, and to give his life as a ransom for many' (Mark 10: 45). When we love people, we are drawn instinctively to help them, to give them service as best we can. But if love is service, sin is the refusal to serve. It is proud, vain, arrogant, self-centred. It will not spend itself generously for the needs of others. Love is knowledge. To love is to possess knowledge of a distinctive quality. Read a newspaper article about a friend or watch a television documentary on a favourite place. Each may have been prepared with evident conscientiousness and care, but you recognize at once that something is missing. It would indeed be unfair to criticize, for only love could supply the lack. And if love is knowledge, then sin spreads ignorance. It is narrow, prejudiced, the seed-bed of moral misinformation. And love causes unity. It binds together those who love. Whatever form of commitment we may think of, personal, professional, national, religious, or any other, the commitment in love generates unity. But sin destroys it. In marriages, in families, between friends, colleagues, nations, sin divides. Time and again, sin is found to be antagonistic to love. It is opposed to everything which is in our best interest.

Furthermore, this love to which we are called is not an abstraction. It is vivid and personal. It means that through our actual fidelity in love, we come to share in the life of Christ. When we are trying to understand the moral life of the Christian, we should see it as a life rooted in Christ, because rooted in love. The consequences can be alarming. They can easily be misunderstood.

Christians are called to share in Christ's life, which will mean a share in his passion and death as well. We have considered already how his passion can become ours. By remaining faithful as he was faithful, we are wounded and at one stage defeated. Some people will be trapped by their social conditions, denied their basic human rights; they may be impoverished or unemployed. Others will bear wounds of a different kind: grave illness or handicap; an unhappy marriage; childlessness or the conception of a child unwanted. The list could be greatly extended. It is important to recognize that the Christian response to such wounds is not a mere shrug of the shoulders, an acceptance that nothing can be done. That glib attitude which assumes that nothing should be done about hardship, but every hardship should be identified as our cross, must be dismissed as beneath contempt. We must struggle to overcome injustice and labour tirelessly to heal people's handicaps and problems. But when all that has been done we will realize sometimes that some situations are insoluble, for the only kind of relief that can be found for them is perceived to be immoral. Then we are wounded indeed, then we are sharing the cross of Christ. The only alternative, championed no doubt by a selfish and materialistic society, is illusory. The relief that is brought proves to be the cause of damage far more terrible than the conditions it sought to overcome. The end can never justify the means, because, when a good end is achieved by immoral means, we can expect only to reap the whirlwind. It is not a question, therefore, of suggesting that nothing can be done. There may be much to be done and sometimes with the happiest results. At the same time, however, we must acknowledge that there are limits to the solutions we can employ and as Christians recognize in those situations that the wounding we receive is not meaningless. We are wounded

through our fidelity in love, as Christ was, and we are confident that
we will find our healing and renewal by remaining faithful to that
love.

Christian discipleship demands our commitment to love.
Through our fidelity to that, our share in the life of Christ is
nurtured, the bond between us is strengthened. He is the head, we
are the body. His love for us has overwhelmed us. By sacrificing
himself for us, that is to say, by coming out of love for us and,
faithful to that love, by dying and rising from the dead, he has
changed our state. By our fidelity to that love, it is possible for us to
be created anew. We carry a cross, but rise again. St Paul describes
those who are in Christ as 'a new creation' (2 Cor. 5: 17). It is heady
talk. What does it mean day by day?

(iv)

At various times in our lives we have to make decisions. We decide
to marry this person rather than that, or we may decide not to marry
at all, perhaps wishing to become a priest or to enter a religious
community. We decide to undertake a particular scheme at work or
to commit ourselves to some fresh project. We may accept promo-
tion. These are random examples, but they have something in
common. When we make such decisions there are consequences for
us. Our lives are changed or more clearly defined by them. Such
decisions supply a direction for our lives or determine their direction
more precisely. And in these circumstances experience soon teaches
us a painful lesson. We may have seen our decision as the solution to
a problem or dilemma. It is generally otherwise. To have chosen a
particular path, however delightedly, does not ensure that we will
walk along it without straying. Our behaviour too often fails to
match our commitment, sometimes in minor matters, sometimes
more seriously. And so the honest will at times compromise their
integrity, even loving husbands and wives may be unfaithful, and
those who are brave may nevertheless be moral cowards at times. A
firm decision, genuine commitment, is no guarantee against falling
short of the ideal, nor does it necessarily cancel the choice we have
made. This is a fact of our experience which we can all recognize,
however ruefully.

It is valuable to notice it here, because our Christian lives—our
decision for Christ—bear the same hallmarks. The choice that we

make should touch the very roots of our being. It does. But we can then be shocked and become unnecessarily despondent, when we realize that this decision has not transformed us instantaneously into perfect Christians. More dangerously, the tendency in the past to categorize our sins too readily as mortal or venial could be deeply confusing and damaging. One mortal sin and a person's whole relationship with God was considered severed until repentance and sacramental confession had taken place. But our decision for Christ, like those other decisions, while it establishes a state of affairs, does not guarantee that we never stray. We have to grow in our chosen life and into our chosen life. Nor will our development be smooth and undeviating. Adultery does not inevitably mean the rejection of the marriage bond. In the same way, our commitment to Christ can withstand very serious failings. We must be trying to grow all the time, sloughing off our sinful choices. It is a long struggle as we try to make our decision for Christ govern everything we do.

(v)

This decision establishes us in Christ, but does not make us perfect at once. All our other decisions should be in accord with this fundamental one. Many of them are. Little by little, we reinforce our basic choice for Christ, sometimes by taking further major decisions in a way that supports it, like entering Christian marriage or the religious life; more usually by the way in which we behave throughout each ordinary day. That is straightforward. A question is raised, however, by those other choices which, because imperfect, we still make, the choices which conflict with our decision for Christ. How are we to understand them? We can refer to our common experience. There are three main possibilities.

First, there are those occasions when we act in ways which conflict with our commitment to Christ, but without destroying it. Although no offence against God is trivial, not all are lethal. Once again, think of marriage. A man may be most happily married, but that does not guarantee that his treatment of his wife will be perfect. He will sometimes be short-tempered, or inconsiderate; he may not always tell the truth; he may be self-centred; he may even by some damaging aberration be unfaithful. But none of these examples, or various others no doubt, need mean that his essential commitment to his marriage is broken. Human inconsistency should not be judged prematurely as rejection. And so it is between ourselves and God. We may sin, and sometimes in fact sin very seriously, but that fundamental decision at the very root of our lives is not thereby destroyed.

Second, there can be occasions when a person draws together his convictions and with full knowledge and deliberate intention changes his allegiance. The event appears sudden and dramatic. It is an experience of conversion, except in this case the decision has been taken to reject Christ, to destroy the bond. Discipleship has been exchanged for opposition to God. These changes of disposition fascinate us, but it is almost impossible to determine their causes. To use the marriage example again, because it is so apt, there are probably cases of married people (although I have never heard of one), who have suddenly decided one day that their marriage was dead and have walked out never to return. They have not been having an affair; there has been no reason to expect their departure. They have suddenly looked at their partner, decided that love has died, and left. And so we may suddenly see life differently and give ourselves up to sin of the most serious kind. We spurn the worship of God, betray those who love us and whom we love, deceive those who trust us, neglect those who had come to rely on us. We choose a life based on self-interest, which is the vainest kind of idolatry. But it is a matter of dispute whether anyone has ever managed to concentrate their disposition in this way and, with one unheralded stroke, repudiate what previously they had prized above everything else. One weakness of moral teaching which emphasizes too exclusively actions which are performed, while neglecting circumstances and intentions, is its readiness to assume so total a rejection of God based on an action deemed mortal sin. Thus a generally devout Catholic, who once missed mass on Sunday, but was then killed in a car crash going to work the next day, would be considered condemned to hell for all eternity. It obviously will not do. Profound commitments are not reversed by chance.

The third possibility is far more familiar. In this case, too, there has been a decision to follow Christ and, as always happens, because we all sin, the person has failed to be perfectly faithful to his decision. Here, however, these lapses have been left unattended. And so a sequence of neglected failure has become a matter of habitual betrayal. These betrayals have taken their toll. Initial faults, repented and wrestled with, bring our lives more and more into harmony with the choice we have made. But the same faults, perhaps slight in themselves, when ignored or accepted, gradually erode the chosen foundation. Here once more the preoccupation with the categories of mortal and venial sin is found to be misplaced, for a person may never commit what might be judged formally to be a mortal sin, but he may nevertheless so persist in his meanness towards others that his relationship with God is destroyed. It is like a marriage that dies, not because of some major infidelity but as the victim of a thousand acts of neglect.

Our decision for Christ is real, but it has also to be made real by our bringing our lives more and more into conformity with the fundamental decision. An American friend of mine once put the matter succinctly, if rather abstractly. He said: 'The Christian indicative is followed by the Christian imperative.' It is not difficult to see what he meant. Thumb through the New Testament and you will find passage after passage proclaiming the good news, setting out, in other words, what Jesus has done for us. And what Jesus has done, you might conclude, has been done definitively. Nothing further is required. But that is not the case. These statements of achievement are coupled with commands to live in accordance with what has been achieved. The starkest example is to be found in the second letter to the Corinthians. It comes immediately after that heady reference to the new creation of all those who have been established in Christ. Paul goes on: 'God was in Christ reconciling the world to himself, not counting their trespasses against them, and entrusting to us the message of reconciliation.' Here is a clear statement of the reconciliation which has taken place through Christ's work. If we have been reconciled, then we are reconciled. Nothing more needs to be added. But Paul at once adds that he is therefore an ambassador for Christ, making his appeal, and his appeal is: 'be reconciled to God' (2 Cor. 5: 19–20). Why do we need to be reconciled, if reconciliation has already taken place? We realize the answer easily enough, for we can all acknowledge a gap between what we are already, in one sense, which is what we ought to be, and what we are in fact. A person appointed to a position of responsibility holds that office in fact, but he must also act in a manner that corresponds to it; from time to time he fails. As Christians, we are in Christ already, but in another sense we have yet to become what we are. We do so by means of each decision we take, the few which mightily support the basic decision and the many which confirm it in a supplementary way. And at the same time we have to wrestle with those which are in conflict with it, for fear that what we have chosen may be eroded.

(vi)

This chapter may have been a disappointment to some readers, if they have been looking for a clear guide to Christian living. Such guides do not exist. I have, moreover, thought it wise to avoid various controversial moral issues, which could easily have distracted us from our principal concern. It has been my purpose instead to underline the primacy of love for the Christian. Fidelity in love brings us to a real share in the life of Christ. Sin by contrast attacks that fidelity in the hope of destroying love. As we grow in Christ, we choose more and more consistently to live in love; accordingly we try to overcome our faults, our sins, which clash with that love. So far so good. But a person may still ask reasonably whether some indications are not available to help us assess the moral quality of our lives. I think they are.

In the past, as we have seen, the principal emphasis was placed upon what was done, the act itself. This approach was intended to safeguard the objectivity of morality: what is right, is always right; what is wrong, is always wrong. Departure from this firm standpoint was thought to open the flood-gates to subjectivity. But it is important to remember the commonplace truth that morality will always involve a relationship between the subjective and the objective. An absolutely objective morality would be cut off from human society: it could have no relevance; while absolute subjectivity cannot be a basis for any morality proper: a free-for-all does not qualify. Moreover, as we have also seen, a preoccupation with the act can be misleading, every 'mortal sin' receiving the same status, be it murder, adultery, or a missed mass one Sunday. We need more sensitive means of discernment.

Once these qualifications have been noticed, however, we should realize that the act itself is a matter of importance. There is generally something more loving about an act of sexual intercourse than a casual peck on the cheek, something more evil about vicious slander than a snide remark. Of course, sexual intercourse can be loveless, while a slight gesture may express the deepest bond. And, of course, slander can be so extreme as to be laughable, while a passing comment can be deeply wounding. But in general the act is a guide to what is taking place.

Intention also is highly significant. The good or evil of what I do, follow from my intention to a large degree. It might seem obviously wrong to slap someone's face, but if they are having a fit of hysterics, that might be the kindest action to take. Intention is vital. The desire to do good should govern our behaviour. On the other hand, it cannot be the only aspect that merits consideration. The vandal who destroys property 'for a laugh' is not blameless simply because he had not thought about the consequences of the damage he has caused and, as he pleads, 'did not mean to do any harm'. Were intention the only factor determining morality, it could be used to justify great evil. Life is evidently more complex than that.

The other main factor which must be considered is circumstance. We should deplore violence. But when we know that the perpetrator of some violent act has been submitted to extreme provocation, we view the matter differently. It may not justify what he has done, but the moral judgement to be passed will be affected by the circumstances. Again, it is wrong to steal, but if a man steals on behalf of his starving family, he is not only excused, but the Church teaches that he is in the right. Their extreme need takes priority over the right to private property.

These comments are necessarily brief, but they may at least serve to introduce a highly complex area. Moral theologians debate the order of priority between these three factors, the act, intention, and circumstances. It is not a debate that can be settled here. I am not myself sure that any such order can be agreed. It seems rather that each one contributes to the moral quality of our behaviour. They relate to one another and we can determine the morality of what is happening only by being sensitive to the whole event.

(vii)

That last sentiment may seem unbelievably bland. More seriously, it leads us on to remember the importance of conscience. Conscience is not some peculiar organ or part of the brain which can be singled out. It is rather a way of speaking of the very core of our being, that centre where we are most truly ourselves. There our moral sense should always be on the alert. We will take care to educate and inform our conscience for that purpose.

To live according to conscience does not mean that each one can do whatever he or she thinks best, nor does it mean to give automatic unquestioning acceptance to everything the Church's magisterium proclaims. Once again, two attitudes must be held simultaneously: first, utter confidence in the Church as a guide in these matters, and second, the awareness that it can rarely comprehend an event in its total concreteness. Thus there will always be a need for communication between the teaching Church and the consciences of individuals and local communities. Each informs the other and, by doing so, ministers to the other, enabling all to grow more perfectly into Christ. When we realize this fact, we uncover one of the great popular errors about conscience. Many people suppose that conscience can be left idle most of the time. Require them to do something they dislike, and they stir their consciences to protest. But conscience does not behave like that. We should not live by conscience occasionally, which will often mean only when we wish to avoid some unwanted duty. Conscience should always be on the alert. Usually and, it is to be hoped, always it will find itself in perfect agreement with what the Church teaches. Why be a Christian otherwise? But that very conscience will mean that, should it ever be stirred to dissent, it is a voice worthy of attention.

Sensitivity to the whole event, however, means more than an alert conscience. It calls for as complete a life in Christ as we can fashion. That life will be sealed with the virtues of faith and hope and love.

II

Living faith

(i)

JESUS was faithful. If one truth has emerged from these reflections, it should be that. Out of love for his Father, the Son became man. He was born, lived, died, and rose from the dead to glory. We speak of his perfect obedience to the Father's will. And this love and obedience were a delight for him, for through them he also showed his love for all men and women. He came to save us not only because he was sent, but because he loves us. Everything the Son did gave expression to that love of his for the Father and for us. Everything he did bore witness to his interior disposition of perfect fidelity to that love. It is not just that he loved; he was utterly faithful to the love. We have seen already that his passion and death, on the one hand, and his resurrection, on the other, were not like events in a programme which had to be fulfilled, but were rather the inescapable consequences and outward manifestation of his interior disposition of obedience and love to which he was completely faithful. We can say that to know Jesus is to understand what being faithful means. And Christians are called not only to know him, but moreover to share in his life. It is not enough for us to understand fidelity, we have to be faithful as he was faithful. Faith is, then, first of all, a way of speaking about our relationship with God in Christ. It is a sharing in that personal attitude and response, fundamental to Jesus, which were characterized by his obedience and love. It is a relationship within which we too are summoned to love and obey. That summons is fulfilled only gradually. It is one of the aspects of having faith which we can find particularly difficult.

(ii)

I cannot help smiling whenever I remember the story of a teenage girl at one of England's more expensive convent schools. Some years ago she went up to one of the nuns who taught her. 'Mother,' she announced boldly, 'I've lost my faith.' 'My dear,' came back the gentle reply, 'you never had it.'

But what makes me smile can make others alarmed. I have told that story before and have had parents say to me, 'But that can't be true. If she was baptized, then she must have had the faith. Don't we receive the gift of faith through baptism?' 'Yes,' I tell them, 'we do.' 'Well, then.'

Part of the problem here is the term, 'gift'. Its difficulty was brought home to me one evening in the late seventies when I was talking to a friend of mine who wanted to become a Catholic. Her desire for Catholicism was strong, but she had no confidence that she really had faith. I may have been very dim that evening. For a long time I could not see what was troubling her. In due course I felt I had to say to her that, if she really did not believe, then, in spite of her desire, it would be better for her not to be received into the Church. My words, of course, only intensified her unhappiness. I left the room and went downstairs to the chapel for ten or fifteen minutes. I could not claim to have prayed in any formal sense, but I was there. The conversation in my room had become deadlocked. I returned as perplexed as when I had left. But as we began to talk again, the difficulty began to dissolve. My friend was disturbed, I realized, because she regarded faith as a gift, something which, once given, was possessed. It was that kind of faith which she felt she lacked. And faith is a gift, but not like a wrapped package which is ours, whole and entire, when it has been handed over. It is a gift as love is a gift. In the first place, it is undeserved. And then, when we receive it, although it is ours, it is nevertheless incomplete. It has to grow and develop throughout our lives. It is properly complete only at the end. Once my friend had come to see that, the last major obstacle to her becoming a Catholic had been removed. She was received a short while afterwards.

The same problem can emerge in another form. Parents sometimes have reservations about baptizing their newly-born children. We referred to some aspects of this viewpoint when we were considering baptism itself earlier. In the present instance, however, there can be a feeling that to baptize will constrain freedom. Interestingly, it is the outlook we met before in those parents who were alarmed by the nun's reply to her pupil. Both sets of parents regard baptism as bestowing the gift of faith. They suppose that once you are baptized, you have got the package, whether you like it or not. One set is pleased, the other apprehensive, but their view of the matter is the same. Their view is wrong.

Faith is a gift, bestowed in baptism, but not a gift like a package. In the first instance, the gift is a relationship. The relationship is real, it is established, but its development and outcome cannot be guaranteed. You may know parents who have loved their children generously and without reserve, only to find that their children later reject them and cut themselves off. Their parents love them still. They possess their parents' love as much as they ever did. But they have no wish to respond to that love, to live in it. It is much the same in our relationship with God. When we are baptized, we give expression to God's love for us. As we have seen, such expressions are creative. They establish us in that love. But the outcome is not guaranteed. God's love for us does not change, but we may grow in it or reject it. The life of faith may be strengthened or severed. Nothing is fixed, nothing determined.

It might be helpful to mention one further aspect of faith as a gift. You may remember that God's love for everyone was emphasized in our discussion of baptism. God does not love only the baptized. The reason for being baptized is not in order to enter the charmed circle of the beloved. Why then be baptized? We should be baptized because, in doing so, we answer that desire, deep within us, to give expression to the love that overwhelms us. The distinction is not between the loved and the unloved, but between those who know they are loved and those who do not. That knowledge is precious. So the question rises insistently: as faith is God's free gift, why does he not give it to everyone? Why do some people believe so easily, and others only with difficulty or not at all?

I must disappoint you. I raise the question to give it air, not to answer it, for I do not know the answer. I have sat for hours talking with people who have wanted to believe, but with utmost honesty have had to declare in the end that they found it impossible. I have seen people grappling with faith, like the friend whom I mentioned earlier, and then suddenly the pieces have fitted together. My own father became a Catholic shortly after marrying my mother. She tells the story of their sitting together on the front at Parkgate beside the river Dee before they were engaged. They were talking about Catholicism and my father was worrying over the whole question of whether to become a Catholic or not. After a long discussion, there was silence. Then my father suddenly said, 'Oh, of course, I see it now. If you have faith, everything falls into place.' And from then on, he was in no doubt about what he would do. And then again, I know people whose generosity and spirit put many a Catholic to shame, but for whom religion and faith in God mean nothing at all. It is a great mystery why faith should be given to some and not to others. But one conclusion should be clear: those who believe must bear witness to their faith by what they say and do. It is a gift for others, to be shared, not a treasure to be protected with miserly vigilance.

(iii)

One of the standard criticisms of faith looks upon it as a form of special pleading, and so as dishonest. How, it asks, is it possible to believe in something which cannot be proved or cannot be understood? In opposition it argues for an approach which will affirm nothing without the support of evidence which can be checked thoroughly. This basic standpoint can be developed in a most sophisticated way. It is another of those areas which have given rise to a vast range of study. It would not be possible to survey it here. I mention it, however, in order to draw attention to our instinct for faith. By instinct, I mean that, if we consult our experience, we find that it is part of our condition. Believing comes to us naturally.

There is a tendency, implied by the criticism, to categorize believing as exclusively religious behaviour, but in fact it is far wider than that. We must believe in order to achieve anything worthwhile. Its lowest manifestation is perhaps our need to hypothesize. Whenever we try to learn something, there comes a time when we have to trust in what we have learnt so far and make use of it. We know it is not exhaustive. We know that there is further evidence untapped, but to learn it is necessary to begin. The first year student who has to produce an essay on *Othello* may have read his booklist with great care and taken well-observed notes, but then he has to write. If his reading has taught him nothing else, it should have made him realize how little he knows. At least his limitations are clear to him. But he must write. So what does he do? He makes a judgement based on the little knowledge he has, chooses a line of approach, trusts in it, and proceeds. He will make mistakes and be criticized, but he will be learning. Without the hypothesis, without that primitive act of faith, he will get nowhere. And academic life does from time to time give us examples of people who are so concerned to check every detail that they stagnate; they produce nothing at all. There is an act of faith in learning. There is an act of faith in living as well. Imagine a lawyer who always wanted to check another authority, a doctor who always asked for a further opinion, a businessman who would never act until the next fluctuation of the market: they would soon be without clients, patients, and prosperity. To live we must act, and action requires judgements based on limited evidence. Those judgements, from another point of view, are acts of faith. Believing comes to us naturally. It is common to us all. And what is true for learning and for life is also manifestly true for our relationships. You meet someone and find you like them. You meet again. You talk. The conversation becomes personal and very soon you find yourself drawn—it is the very word we use—drawn to *confide* in them. And sometimes at the last moment you draw back. Perhaps the confidence is shared at a later meeting when knowledge has grown, but you are not prepared to reveal yourself so early. Sometimes the instinct for reserve is wise, as the friendship proves to be fleeting. On other occasions, however, the rapport between new friends can be so immediate that they will open up to each other without hesitation. They scarcely know each other, but their faith in each other is strong and leads on to depth of friendship, while the person who is always reluctant to be trusting, is destined always to be lonely. Believing is not only natural for us. It is also necessary. It is by no means the exclusive preserve of the religious. What we have to consider is its significance within the religious setting of Catholic Christianity.

(iv)

From one point of view, faith is a codeword. It is our way of speaking about our relationship with God. The bond between us is not our due, it is a gift to be nurtured and developed throughout life. It is a personal relationship and it is an attitude which enables us to share the life of Christ. To be filled with faith, faithful as he was faithful, is to possess that interior disposition which, we have seen, shaped his entire existence. Through our faith we are at one with him. Such a conclusion follows naturally from what we have considered before. Faith is the basis of our life in Christ.

Faith is necessary for learning as well. The connection between self-knowledge and identity has been another constant theme of these reflections. More particularly, when discussing the risen Christ, I remarked that he was 'truly present, but visible only through the perceptiveness which faith bestows' (p. 33 above). He was seen, risen from the dead, by faith. And indeed the point is more general. These reflections have taken their cue consistently from the belief in Jesus as true God and true man. They have insisted that the divinity which is distinct, which has maintained its integrity, is nonetheless present to us in and through the human. It does not display additional credentials. And so the reality of the divine has to be discerned within the human. We must examine this view more carefully. As it stands, it may well seem to illustrate that form of special pleading, to which I was just referring, a special pleading which can so swiftly arouse the suspicions of the non-believer.

From one viewpoint, however, the perceptiveness of faith resembles an experience with which we are all familiar. Call to mind something very ordinary. A mother asks her football-crazed young son where he has been playing that afternoon. She expects an answer of his friend's house or garden or down in the park, but the instant reply is 'right wing'. Or again, we are familiar with the phenomenon known as 'naming the beloved'. It is the tendency of those in love to introduce into almost any conversation the name of the one they love. They do so virtually unawares. In other words, in ordinary life we find that, when people have some absorbing interest or some deep-rooted experience—perhaps of love, but it could sadly be of hate, it could be a sorrow or a joy—whatever it is, it affects the way they see their lives. This kind of interest or experience has given them a fresh perspective. It may be regarded as extra evidence; more usually it might be said to throw light on the evidence and so unveil it more completely. And what we observe in an ordinary way takes place in Christianity through faith when we consider our relationship with God.

In his first letter to the Corinthians, St Paul spoke of the wisdom which was imparted through the Spirit. These are his words: 'Now we have received not the spirit of the world, but the Spirit which is from God, that we might understand the gifts bestowed on us by God. And we impart this in words not taught by human wisdom but taught by the Spirit' (1 Cor. 2: 12–13). What he was saying repays attention. The Spirit that he has received is the Spirit of God, not the spirit of the world, and he has received that Spirit for a purpose: so that he might understand the gifts bestowed on him by God. The

Spirit enlightens him. This enlightenment, this understanding of God's gifts, is itself to be made known, but, Paul continued, not in the words proper to human wisdom; it is to be imparted by words taught by the Spirit; spiritual truths will be interpreted to those who possess the Spirit. You can see what is being described here. Paul has presented a sketch according to which the mysteries of faith are recognized and made known within the setting of the Spirit's action. The point is crucial. He is describing a whole process of learning through the Spirit. As there is a perceptiveness appropriate to logic and music and art, to the intellect and to the emotions, so there is a perceptiveness appropriate to faith. 'The religious mind', as Cardinal Newman once observed, 'sees much which is invisible to the irreligious mind. They have not the same evidence before them' (quoted in Wilfrid Ward, *The Life of John Henry Cardinal Newman* II (London, 1912), p. 247). There is nothing sinister about that claim. It is the predictable consequence of our sharing in the life of the Christ.

This perceptiveness which faith bestows is a matter of vital importance. I have referred already to the way it enabled the apostles to see the risen Lord. There are many other examples of the part it has played in the life of the Church. It is essential for that self-knowledge which preserves the identity of the community. I have mentioned it on various occasions. It is this perceptiveness which has enabled the Church to recognize the divinity of the Christ, to know itself as the people of God, to decide which documents articulated its faith so profoundly as to be acknowledged as divine revelation, to judge which teachings should have the status of dogma and which acts were sacraments, acts of divine love in our midst. It is a perceptiveness born of a life rooted in faith. Of course, if you believe that our divinely revealed religion was established by blueprint from on high, such considerations will seem quite by the way. But at the heart of these reflections has been the belief that God became man in Jesus of Nazareth. It was not a charade, he did not adopt a disguise. The divine Word of God became a man. The divine and the human were united perfectly in him and the divine was revealed by means of the human. This realization that faith is perceptive is vital for us. It means that, as we share in the life of Christ, in other words, as we are faithful as he was faithful, so does that life of faith bring us the light to recognize the divine presence within the human. To affirm that we hold our convictions by faith is not to indulge in special pleading, but to acknowledge that process of perception which is appropriate to faith.

Sceptics will have a question. If faith brings understanding, they will want to know, why are Christians so divided? Faith is disposition, not just content of belief. Are there not sincere believers in every Christian tradition, men and women of undoubted holiness? How can they be so opposed to one another?

A part of the answer may lie in a misunderstanding implicit in the question. Many who are sceptical about Christianity presume that it can make no allowance for diversity of view. They expect complete and perfect uniformity. But Christianity is not like that. It has always recognized and respected a variety of interpretations, conscious that no single human expression of divine truth could be exhaustive. Some generations may have narrowed the variety more than others, but the principle has been beyond serious dispute.

Another part of the answer is more challenging. To put it simply, the divisions between those of us who believe so sincerely, bear witness to the immaturity of our faith. The disagreements show up the limitations of our life in Christ. It is no solution to forget the disagreements and place emphasis on that life alone. That strategy would be based on illusion. The disagreements exist because the life of faith is flawed. Ignore them and we will soon be imagining a depth of common faith which is in fact illusory. A deep and absorbing union will be one of mind as well as heart. Allowing always for the possibility of rare exceptions, a strongly loving married couple are most unlikely to combine their relationship with bitter disputes over matters of conviction. Either their love will lead them to agreement, or else their clashes will erode their love. If Christians are to achieve the unity for which Christ prayed, their first priority must be their own lives of faith. As we grow in living faith, our understanding will also mature: we will slough off the skin which might once have seemed so essential and recognize the value in what had once seemed so inadequate. We will be like musicians who have dismissed a score as cacophony, only to discover that a more developed musical sensibility and more finely tuned instruments reveal a glorious harmony. As our life in Christ matures, our divisions will be overcome.

(v)

Faith is not only a way of speaking of our relationship with God and of our perceiving the mysteries he has revealed; it also has a content. To have faith in Christ means to hold particular beliefs. These pages are attempting partly to give an account of those beliefs, beliefs about Christ and God, about the Church, the sacraments, and the Christian life as it is made real through the virtues of faith, hope, and charity. The scope is vast. We know that such an account must be incomplete; it cannot include everything, however objective we may try to be. And so the approach is evidently personal. In the first place, I felt that to reflect and draw upon my own experience of Christian faith and life was the best way to select what needed to be said. And in the second place, there may have been a further advantage. This approach may have helped to illustrate the inter-relating of the Catholic tradition of Christianity and personal response. Of course, if all I have managed to do is give an account of Roderick Strange's faith of which readers may approve or dis-approve, in which they may be interested or by which they may be bored, but with which they can find nothing in common, then I have failed, for the personal approach will have obscured what is authen-tic itself and has been given for all. Faith has a content. There are beliefs to which we adhere. But if through this approach others can recognize their faith or at least the possibility of their believing, then I hope that is because they can recognize something which is not just personal to me, while in the personal they may be discovering clues to the character of their own personal response.

This decision to believe is a curious experience. Any priest who has sat for hours talking with those who are wondering whether to become Christians or, if baptized already, Catholics, knows what an unpredictable process it is. For some it is easy, for some difficult, for some impossible. But when the decision is taken, when people make an act of faith, there is something absolute about it.

I know that philosophers debate the question but the pastoral reality I have seen often. People do not believe up to a point; either they believe or they do not. The act of faith is quite distinct from the arguments upon which it is based. Good arguments may convince, but not necessarily; and sometimes convictions remain firm, although the supporting argument is demonstrably poor. The situa-tion may apparently be even more extreme. I remember the philos-opher, Michael Dummett, being interviewed on television once. He was asked about his conversion to Catholicism which had happened many years previously. He replied that he could no longer remem-ber the arguments which had swayed him, but doubted very much whether they would have moved him still. Perhaps he was implying that he now had a fresh and more satisfactory set of arguments, but I understood him to mean that his faith did not depend upon the arguments associated with it; they had faded, but the faith remained. Pastoral experience will often bear this out, usually with the circumstances in reverse. In other words, I have known people resist faith because they were so aware of the difficulties which believing raised. They have felt it was dishonest to profess a faith when they could still see problems in it. As they have expressed it to themselves and to me, they have had their doubts. So I have suggested to them that there is a distinction between difficulties, the problems which may surround any position, even a belief, and a doubt which is directly in conflict with faith and incompatible with it. And with the explanation the cloud has often lifted. They have recognized what they had called their doubts as queries about belief, not as doubts properly speaking. The problematical arguments need not compromise the act of faith. A life of faith has been accepted and undertaken.

To acknowledge a distinction between the act of faith and the evidence on which it rests, does not imply that the evidence itself is unimportant. The customary mistake here has been to look for evidence of the wrong kind. People have sometimes expected a formal, rational argument, something logical, and so, as they have thought, watertight. It will be better to return to an earlier point, the perceptiveness faith bestows. And as there is a perceptiveness appropriate to faith, so must there be evidence which corresponds appropriately to the perceptiveness. We do not read with our ears or listen with our eyes. Our senses need the proper kind of experience to communicate to us. Our religious sense is the same. We cannot expect to penetrate the mysteries of faith with abstract logic. We have to assemble the evidence from scripture, tradition, the teaching of the Church, our own experience of faith, of prayer, and so on. Little by little we may acquire a picture of the faith which, we find, rings true. We are convinced of it. Cardinal Newman once used the image of a cable to express this idea. It is, he said, 'made up of a number of separate threads, each feeble, yet together as sufficient as an iron rod' (eds. C. S. Dessain and E. E. Kelly, *The Letters and Diaries of John Henry Newman* XXI (London 1971), p. 146).

We have considered faith so far as our personal relationship with God. We have seen that it is not only the relationship, but a way of perceiving the relationship and moreover that that perceptiveness unveils God's truth for us. It would be reasonable to suppose that this friendship with God and this knowledge of his ways would be sources of intense comfort and delight. They can be, but we must not count on it.

(vi)

What does it mean, to live by faith? What does steadfast loyalty to God bring us? My fascination with the people's experience as presented in Psalm 44 (43) never fades:

> We heard with our own ears, O God,
> our fathers have told us the story
> of the things you did in their days,
> you yourself, in days long ago.

The psalm begins by recalling God's goodness to his people, the blessings he has poured down on them. But after the memory of praise, the mood changes:

> Yet now you have rejected us, disgraced us:
> you no longer go forth with our armies.

The psalmist recognizes that God has turned away from his people. If we are familiar with this literature, we should now expect an acknowledgement of sin and a plea for forgiveness, but that expectation is not fulfilled. Instead there is a cry of bewilderment, for God's displeasure seems to be inexplicable. The people have not forgotten him, have not been false to his covenant, have not withdrawn their hearts, nor strayed from his path. The psalm ends with a plea to the Lord to redeem his people from their misery and oppression because of his love for them. The people are in crisis and God appears to have deserted them. Their fidelity to him does not seem to have ensured his action on their behalf.

In one of his sonnets, Gerard Manley Hopkins expressed the same experience:

> Wert thou my enemy, O thou my friend,
> How wouldst thou worse, I wonder, than thou dost
> Defeat, thwart me? Oh, the sots and thralls of lust
> Do in spare hours more thrive than I that spend,
> Sir, life upon thy cause . . . (*Poems*, no. 74)

It is the prosperity of the wicked even in their 'spare hours' which I always find so affecting. The life of faith is no guarantee of ease. So what is it meant to be? Are believers supposed just to hang on blindly when troubles come? Are they just to cling to Jesus, though almost paralysed with fear? We feel somehow there ought to be another explanation. Indeed, when the apostles were afraid, Jesus rebuked them for their lack of faith.

They were crossing a lake. A storm blew up and they panicked. All the while Jesus was asleep. So they awakened him. He questioned their fear, 'Why are you afraid?' And then he asked, 'Have you no faith?' (Mark 4: 36–40). The point is plain enough. If they had had faith, there would still have been a storm. Their faith would not have been a shelter against crisis. Jesus, the faithful one, was in the midst of the storm as well. But if they had had faith, they would have reacted to the storm differently.

We are back to the familiar refrain. God's people are called to be faithful as Jesus was faithful. His faithful obedience to the Father's will and love for the Father and for us did not protect him from conflict. On the contrary, as we have noted time and again, it set him on a collision course with a world scarred by sin. But because his fidelity never wavered, his defeat on the cross did not end in death; he overcame death and rose to new life.

To live by faith is no shelter from defeat, but rather a way of facing and suffering defeat. If we allow the defeat to smother in us a generous loving heart, then our defeat is a defeat indeed, death-dealing. But if we are faithful in love, obedient to the Father, so far as lies in our power, and unfailingly generous in our love for him and one another, then we pass on from defeat and enjoy new life. We discover within ourselves the death and resurrection of the Christ, stamped on our hearts like a seal. Through this living faith do we share in his life.

This faith brings forth fruit in perfect love. Love overwhelmed Jesus, it possessed him. It must also overwhelm and possess us.

The command to love

(i)

ONE day Jesus was leaving Bethany. He felt hungry and, seeing a fig tree, he went over to it to pick some figs. But it was not the season for figs; he found nothing but leaves. He said to the tree, 'May no one ever eat fruit from you again.' The following morning, when they passed the tree, the disciples saw that it had withered to its roots. It may seem to us a strangely petulant incident. We should read on. It is used by Jesus as an opportunity to impress upon his followers the power of faith. 'Have faith in God,' he tells them. 'Truly, I say to you, whoever says to this mountain, "Be taken up and cast into the sea," and does not doubt in his heart, but believes that what he says will come to pass, it will be done for him' (Mark 11: 12–14, 20–3). I mention the incident here because of the image of faith moving mountains, for St Paul tells us that if we have such faith, but without love, we are nothing (1 Cor. 13: 2). We are brought back once more to the centrality of love for Christian life.

Jesus was moved by love. He loved his Father and loved all of us. Time and again, we have noted his fidelity to that love. To love was his great command to us through which every other commandment is to be fulfilled. In the sacraments God shows his love for us and our morality is based upon our commitment to love. We Christians believe that we have been invited to share in the life of Christ. We do so not by mere imitation, that is to say, not by copying the behaviour of someone who is at a distance from us, but by participating in the inner reality of his life, by being faithful in love as he was faithful. Love is the source and substance of our calling, and it is our destiny.

We speak of love so often. Let me try to say something more specific about the character of Christian loving.

(ii)

It will be helpful to recall the ideas which we considered earlier when we were introducing the sacraments. We men and women are physical and spiritual beings. That is our condition. What is deepest and most interior seeks outward manifestation. That outward sign is not to be identified with the interior reality, but neither is it to be altogether distinguished from it; rather, the sign or symbol participates in what it represents and is creative of it. In the sacraments God is loving us in a way that corresponds to our condition. We receive his limitless love through these particular acts. All this should be familiar to us by now (see above, pp. 81–4).

And we ourselves, who are commanded to love, fulfil that command by loving particular people in particular places at particular times. We are sceptical when people claim that they love others in general, for we suspect that they seek thereby to absolve themselves from caring for those immediately about them. But charity, as we say, begins at home. That always seems to me a telling phrase. I understand it to mean not that love in the family is the first and easiest step in loving. There is an obvious sense in which that may be true. More profoundly, however, I take it as recognition that love for those most closely and constantly about us can be extremely demanding. Most of us can manage to be charming for an evening or a weekend. Love requires much more. The family is a genuine school for love. Love there and you will really have learnt about love. And love is infectious. When families are loving, they tend to be part of a circle of loyal friends. Our loving can never be an abstraction. Moreover, those who are fortunate enough to have close friends can more easily develop a wider generosity to all whom they meet, whether as acquaintances, colleagues, or by chance. Priests are well placed to know the truth of that observation.

In the past they have sometimes been warned against friendship. A cheerful detachment was encouraged instead. Everyone was to be treated in the same way. The warning was intended partly to protect their celibate state, but partly also to help them avoid arousing jealousy amongst their parishioners. Those who ignored the warning sometimes found in fact that some parishioners remarked jealously that their priest seemed closer to some people than to others. It is to be hoped that they were right, but their jealousy was as misplaced as the warning. We cannot be equally close to all. Only if we are genuinely close to some will we be able to give to the others the generosity and care for which they are looking. Those treasured friendships make possible that service. Forbid them and you disable us. When priests become strangers to love, their ministry will soon be only a hollow formality. No one should be such a stranger. We must consider our loving more carefully, particularly with regard to marriage.

(iii)

What has been said about our loving being particular is true outstandingly of marriage. In this sacrament Christians celebrate the pledge which one man and one woman make to each other, a pledge which binds them in an indissoluble union for the rest of their lives. But this exclusive love should never be seen in competition with the inclusive love of the great commandment. Part of the excellence of marriage derives from its capacity to make people love more generally and more generously. Some happily married couples, of course, may keep very much to themselves, but those families with devoted friends will normally have their roots in contented marriages. Such love is attractive.

Loving is also a source of knowledge. By loving we grow in that self-knowledge which is vital for our identity. It reveals to us who we are. Family, friends, every relationship can be instructive, but the deepest loving is a revelation of the couple to themselves. That is the kind of love upon which a Christian marriage should be based.

Those drawn into such a relationship find in the other not just someone who is attractive; they find more than a companion, more than a friend. To be drawn into this love is to be taken by surprise, even to be shocked. That will be true particularly when people have always seen themselves as fairly self-possessed, for the first effect of this love will have been to reveal to them what they lack. They may have been confident that they were whole. Now they know that that was a mistake. A young wife, writing to her husband from London during the blitz, expressed vividly what I have in mind: 'Sometimes in London I look up at the raw edge of masonry where a room has vanished from the other room, and I feel that I know that loss and incompleteness as well as I know anything in life. I'm not a whole person alone, and the edge of the tear hurts all the time' (Jill Furse, quoted in Laurence Whistler, *The Initials in the Heart*, 3rd edn (London, 1975), p. 173). When we love like that, we discover in the other not something extra, to be added on to ourselves, but someone who is part of our very self. The other reveals to us who we are. The sentimental old cliché about our 'better half' is suddenly an obvious truth. It is a matter, not of complementarity, but of completeness.

I often wonder, when marriages fail, whether that is not a consequence of a low expectation of love. People will say that they love the person they marry, but the question remains: have they hurried to identify as this profound love a relationship which may be worthy enough in its way, but which lacks true depth? Have they settled for complementarity without waiting for completeness?

Understood in such terms, this love which makes us whole, may appear so perfect as to grant lovers immunity from all possible problems. Of course, it will do nothing of the kind. Let us consider two such problems.

First, the alternative partner. To say that love is revealing does not mean that the happily married have found in each other the one and only person whom they could ever have married. In other words, there are alternatives. Consider the facts. One of the most encouraging features of my life as a priest has been to realize, in spite of so much writing to the contrary, that many people are very happily married indeed. They may all have their difficult phases; it would be surprising if they did not; but their roots are firm. Now it often occurs to me that even the most gregarious amongst us has a relatively small circle of friends within which our partner has been found. So it is reasonable to suppose that, had we not met, we might still have married, but someone else, of course, and still been extremely happy. This line of thought is not meant to encourage carelessness in our choice of partner. It would not be wise to be profligate. Some people never find even one. But it may help us to realize that there could be an alternative to the partner we have married and very occasionally it is possible we will recognize such a person in someone whom we meet.

It will probably be a difficult moment. A marriage without a foundation in the love I have tried to describe, may well buckle and break. Even in a strong marriage the experience may be severely testing. It is always hard to turn away from the possibility of love. The experience may resemble nothing so much as bereavement. The sorrow may be intense. We realize that a love, good in itself, is lost to us, for to pursue it would mean betraying the love to whom we are already committed. Furthermore, if this new love has been recognized correctly, if it is a genuine alternative, bearing the seeds of truth within it, then it can have no future in these circumstances. A truthful love cannot be built securely on betrayal. Those seeds will never flower in unfaithful soil. Our sorrow will be mingled with a sense of loss and of waste. It is the kind of situation to which we have often referred; we are defeated for it can never be good to have lost such a love. And if we allow the defeat to embitter us, it can do us irreparable damage. But if we remain faithful in love to the one to whom we are pledged, keeping open and generous hearts, then we will find ourselves little by little passing on from defeat to discover a still greater joy than any we had known before. The familiar pattern of cross and resurrection, Calvary and Easter, will be confirmed once again. The unbearable pain will be borne and then transformed. We reaffirm the love to which we are committed, in which and through which we become whole. Such a love will not protect us from pain, but its summons to fidelity is a priceless gift, for it can save us from losing our identity.

Secondly, infidelity. Does the love I have been speaking about make infidelity impossible? It does not. Our capacity for sin is far-reaching. The most authentic loves can be wounded gravely and even destroyed. We cannot suppose that our ability to be unfaithful to God cannot be matched by our ability to be unfaithful to one another. But when people love so deeply and are yet unfaithful, they commit a terrible wrong against themselves as well as against the other. Where the bond is so profound, infidelity undermines their true identity. Under threat, when likely to betray or be betrayed, we need to go deep once more and reaffirm our decision to love.

This admirable love which forms the basis for Christian marriage cannot guarantee freedom from the sad consequences of our weakness and imperfection. We do not prize it for that reason. But to have some idea of it is a great advantage as we try to appreciate what marriage wishes to make real. And here there is a problem. Although Britain may regard itself as a post-Christian society, it remains deeply influenced by Christianity, and so marriage in a church will often be preferred; it may even be regarded in a vague way as more genuine. This disadvantage is double-edged. On the one hand, this attitude can easily appear to cast aspersions on the civil marriages of those who do not believe and on the religious marriages of non-Christians, and, on the other, it fails to do justice to what is distinctive about Christian marriage. People turn up to be married in church like their mothers, or because it still seems the thing to do, and they can be quite unaware of the awesomeness of what they are accepting.

What the Gospel demands of Christians who marry may well seem daunting. Today more than ever human society is conscious of its changeability; we know that our lives are not fixed or static. Yet Catholics believe that the pledge of indissoluble union which two people make to each other in Christian marriage is the one unchanging point in their constantly changing lives. On the face of it, it is a most improbable undertaking. It is not something we should expect to succeed. As we look across history, we do not find many cultures or societies which have adopted such demanding marriage laws. Nevertheless, what Christians believe about marriage is not just their own chance invention. We would say that when people reflect carefully on the human condition, they could conclude that only in the committed exclusive love of a man and a woman is so deep an experience of loving available. However, it is not readily accessible. We have need of grace or, in other words, the presence of God. That presence, I like to think, is not so much a ghostly third partner; I am always rather repelled when I hear preachers say that Christian marriage takes place between three, husband, wife, and God; rather the divine presence transforms what would otherwise be only human. We pick up again the theme of God made man: the divine is not extra; the divine is distinct, but perfectly and harmoniously at one with the human. Perhaps I may be allowed a private view.

At their weddings couples sometimes like to hear the account of Jesus at the marriage feast of Cana (see John 2: 1–11). In the past that has been identified as the occasion when he instituted marriage as a sacrament. We noticed earlier, when considering sacramental life, that that interpretation can no longer be maintained (pp. 85–6 above). The marriage itself was insignificant; the occasion was noteworthy, because the changing of water into wine was the first of Jesus' signs and these signs give his ministry its unity in the Fourth Gospel. All the same, the reading is fitting for a wedding. Water is taken and transformed. Indeed—and this is my quite idiosyncratic interpretation—I like to think that the contents of the jars had always been wine, but the dull palates of the host and steward had managed to mistake it for ordinary water; they had assumed that what was contained in ordinary water jars could be nothing else. But the presence of Jesus changed them. Their perceptiveness was heightened. They tasted again and realized that what they had taken for water, they had mistaken for water. It was in fact most excellent wine, and they possessed it in abundance. And so in marriage human love is taken and transformed. What had seemed good enough, but ordinary, is transformed by the presence of Christ, or rather bride and groom, if they will permit it, are changed by that presence and become capable of perceiving and experiencing their love with fresh and hitherto unimagined power. The bond that unites them is more magnificent than they had ever dreamed. I do not want to be misunderstood. I am not suggesting that such a love is the exclusive preserve of the Christian, although I would affirm that any profoundly loving bond is the work of grace. That grace is active often amongst those who are not Christians or who do not believe in God at all. Christians believe, however, that the more deliberately their love is rooted in God, the more it is deeply and richly perfected.

By love we are made whole. Our true identity is revealed. We come to know ourselves and, through the presence of Christ, that knowledge touches the very depths of our being.

Nevertheless, whether their view of love is exalted or simple, people rarely need to be convinced of the wonder of love. They desire it, search for it, and delight in it. Love is not the problem. The problem arises from knowing how best to express it.

(iv)

Once again, we must go back to that point which is also essential to the sacraments, that we human beings, who are both spiritual and physical, wish to give outward expression to what is inwardly most vital to us. Our loving is not made perfect in undiluted abstraction; the spirit must become flesh. We need to discover how our loves are best to be expressed. We must not be indiscriminate.

In general, we discriminate most carefully in our relationships. We realize that honesty is vital to us here; without it our integrity will be lost. So we avoid those who gush of their affection for us; we suspect their sincerity. We recoil from the kisses of casual acquaintances and are hurt by the cool or dismissive attitude of those whom we care for deeply. So far so good. The difficulty arises when we try to understand the part our sexuality ought to play in our loving.

We happen to live in a society which gives a high priority to sexual activity. It is readily assumed that everyone has a partner for sex. Those who do not are objects of fun or dismissed as inadequate. It is not surprising when those people feel that they have been left on the fringe. Yet this attitude which is so prevalent, can also be rather confusing. People sometimes detect in it an inconsistency. They wonder how an activity which is placed so high on our list of priorities, can at the same time be encouraged and indulged indiscriminately. We need food and drink, but a balanced diet as well. The discrimination we show in our relationships more generally must also include our sexuality. That concern is the hallmark of the Church's teaching.

I wonder how many Catholics agree. So many detect an inconsistency in the Church too. The point can be illustrated very simply. What would be the reaction to news of another Vatican pronouncement on sexual matters? Delight or depression? Depression. Why? Because, whether fairly or not, people have come to expect the Church's teaching on sexuality to be rigid, severe, authoritarian, negative, and completely out of touch with their needs and experiences. They may find a priest who is sensitive, whose approach is sympathetic and pastoral, and they are grateful for that. All the same, they feel uneasy when the Church's teaching can make sense to them only on those terms. If love is truly at the heart of Christian life, their instinct tells them that it ought to be possible to say something more positive and encouraging about its most absorbing physical expression. There should be no hovering shadow of fear and guilt. People should be able to make love with delight.

It may be easy enough to look back across the history of the Church and judge its teaching on sexuality as repressive and so forget the powerful sexual images which have been used commonly. The marriage of Hosea to his beloved, but frivolously unfaithful wife, is used to express the Lord's relationship with his people. The sixteenth chapter of the Book of Ezekiel takes up the same theme when through the prophet the Lord speaks to Jerusalem:

And when I passed by you, and saw you weltering in your blood, I said to you in your blood, 'Live, and grow up like a plant in the field.' And you grew up and became tall and arrived at full maidenhood; your breasts were formed, and your hair had grown; yet you were naked and bare. When I passed by you again and looked upon you, behold, you were at the age for love; and I spread my skirt over you, and covered your nakedness: yea, I plighted my troth to you and entered into a covenant with you, says the Lord God, and you became mine (Ezek. 16: 6–8).

The passage goes on to speak of God's goodness, the loved one's gross infidelities, and God's promise to be faithful to his promise and forgive. The Song of Solomon draws constantly and powerfully on erotic imagery. The letter to the Ephesians understands the relationship between Christ and the Church in terms of the most devoted married love. Much more could be added. These obvious examples may be enough to illustrate a positive awareness of sexuality within the Church. We should realize also that this imagery has been used and accepted. We may think most naturally of St Teresa of Avila's *Conceptions of the Love of God*, which contained her reflections on the Song of Solomon. There, for example, she wrote: 'Again, my God, I speak to Thee and beg Thee, by the blood of Thy Son, to grant me this favour: "Let Him kiss me with a kiss of His mouth". For what am I, Lord, without Thee? And what am I worth if I am not near Thee? If once I stray from Thy Majesty, be it ever so little, where shall I find myself?' (*Conceptions*, ch. IV). This quotation is brief. It cannot do justice to a glorious sexual quality which pervades the entire work. I suggest that the Church has not been dismissive of our sexuality. It has been keenly aware of it, but too often bewildered in its response to so powerful a force within us.

To do better we must take our cue from our own condition, from that union within us of the spiritual and the physical. We should then recognize that our capacity for loving deeply needs to be expressed. It is good to acknowledge here a certain arbitrariness in the Church's teaching. There is a wide range of loving relationships available to us from marriage to chance meetings in a bus queue. No one should be beyond the scope of our loving, but there is a limit to the physical expressions of love available. There may be no necessary connection which limits sexual intercourse to the married state. The point was put to me forcefully some years ago after I had given a talk at an Oxford college on 'Christian Marriage, Morality, and Relationships'. (My host, who had arranged the title, acknowledged with a smile my ironical suggestion that the topic might be too narrow.) One man asked, 'How can you say that, when I make love to my girlfriend whom I love most sincerely, that is less an act of love and less an experience of love than it would have been, if I were a Christian and married to her?' I answered that I would not presume to judge the quality of his love. Of course, there may be acts of love between those who are not married to each other which express a deep and most genuine love, just as tragically there may be such acts within marriage which are indistinguishable from rape. But I added something further and it is crucial here.

If the Church is right when it perceives the value of an exclusive committed love, right to celebrate and support and protect the vital self-knowledge and wholeness such loving brings, and if the Church is right to recognize the need that we have for what is spiritually most profound within us to find outward expression, then it is not being arbitrary, but showing true wisdom to preserve for married love alone our most intimate and joyous physical expression of love. When a man and a woman recognize that the deepest loving bond unites them, they look instinctively for a way of expressing it physically, which will correspond to the exceptional character of their love. Only great intimacy will satisfy. Should that have become a routine feature of every past emotional link, they will be at a loss. To put it differently, in a world where everybody is called 'darling', what terms of endearment can we use for those whom we truly cherish? The currency of our symbols will be devalued, when we use them without respect for the reality they represent. Our appreciation of them will be dulled. We must care for our symbols. They are creative of that reality. Sexual activity, divorced from the deepest love, can do great personal damage. The evidence is to hand in the love affairs that have foundered when the sexual act revealed to the unsuspecting lovers the superficiality of their relationship. Such experiences cause great hurt. Yet many will still laugh at such a notion. They will point to their own experience as proof and ask what harm has come to them from regarding sexual activity as just a means of sensual satisfaction. And who can tell how their indiscriminate sexual behaviour may have disabled their capacity for the profoundest loving? It may be that the readiness with which people accept more relaxed standards of sexual behaviour bears witness to their low expectation of what the alliance between our deepest loving and its most intimate sexual expression can achieve. But we must recognize that when sexual activity is allied to that deep interior loving, spirit becomes flesh, it has power, it makes it real and alive in our midst.

I have concentrated upon married love so much and upon the place of our sexuality within marriage, because it is there that the great command to love is most vividly fulfilled and because that love is also the practical source for a much broader, more generous loving. But to have said so much about sexuality as expressive and creative of love and as inextricably bound up with our deepest loving must give rise to further questions. These pages cannot be a compendium of Catholic sexual morality, but a comment on two issues seems unavoidable. First, what are the implications of what has been said here for family planning? Secondly, if sexual love should be reserved for those who are most deeply committed to each other, what are the implications if those deeply committed partners should be of the same sex? Some people will tell me that these questions do not need to be raised, because the Church's teaching on them is perfectly clear. But there are those who come to see me, a pastoral priest at their service. They are not blind with malice or hard of heart, but they are bewildered by that teaching. It is not possible to resolve their problems here, but it would be wrong not to notice them and at least attempt to comment on them.

(v)

To understand what the Church teaches about family planning and artificial contraception, it is necessary, first of all, to acknowledge the emphasis placed upon the family. When husband and wife come together, overwhelmed by their union, and make love and when from that joyous love-making which so captures their inner union, a child is conceived, their whole relationship is most wonderfully expressed. It is a nonsense to make the begetting and conceiving of children marginal, a mere by-product of sexuality. It is an essential part of it. At the same time, this procreative aspect of love-making is not to be seen in isolation. It is an expression of love and the Church has given increasing prominence to the act of love as a symbol of married love. Symbols, we have recognized constantly, are creative. It is an outward sign of an inner reality in which it participates. Those who love most deeply long to give expression to their love. And here there comes what so many see as a grave difficulty.

There are those whose commitment to family life is unquestionable and who in obedience to the Church have planned the size of their family with great care. They judge that it would be wrong to extend their family further; it might be very large, it might be very small; that is for them to judge. They have been instructed to be responsible parents, and have acted as wisely as they can. It is plain that they are not victims of that contraceptive mentality against which Pope John Paul II spoke on his visit to Britain in 1982. But they know that the Church also urges them to give expression to their love, to use the symbols which nourish it. Abstention may be good sometimes, but not indefinitely; it would conflict with this teaching. The resolution of this dilemma between the care for the family and responsible parenthood, on the one hand, and the sustaining loving, on the other, seems to be found in contraception. But, as everyone knows, the teaching of the Catholic Church forbids the use of artificial contraceptives, and in the summer of 1984 Pope John Paul also warned against the coldly calculated use of natural methods of family planning (see *The Pope Teaches*, C.T.S., 1984(8), p. 248). There seems to be an impasse at this point.

We live in an imperfect world. It is not clear how these teachings on the importance of the family and responsible parenthood, on the value of sexual love between husband and wife, and on the immorality of contraception can be reconciled in all circumstances. And the situation is made still more complex by the fact of so many women, old as well as young, who are most docile and obedient Catholics, except when it comes to family planning. On this issue they are radicals. If the witness of believers means anything, we cannot disregard the convictions of those whose general loyalty to the Church never wavers. It may be that here too—as with our understanding of Jesus as true God and true man, as with our understanding of our redemption—we need to give proper weight to all the aspects of this teaching and allow them to make their impression deep within us. It is another task for that perceptiveness which faith grants.

More immediately and practically, however, it may be that the Pope's words in the summer of 1984 provide a clue. The secular press was rather scornful. Journalists pitied the poor Roman Catholics, who, already forbidden the use of artificial contraceptives, were now finding access to natural family planning restricted as well. But the Pope in fact had fastened on an inconsistency often noticed in the past, namely that it is possible to use natural means of family planning as immorally as artificial contraceptives. What the two would have in common is that negative attitude to family planning which at York in 1982 he called the contraceptive mentality. It may be, therefore, that a positive attitude, marked by purity of heart, could help most to resolve the impasse. I do not suppose that the Pope was meaning to reduce his teaching on family planning to a matter of right disposition. There is plainly much more than that to be taken into consideration. Nevertheless, that point may help to overcome the impasse and indicate the way ahead.

Homosexuality raises other complicated questions. Love is at the heart of Christian life; and our particular loves, as we have seen, enable us to fulfil the Lord's command to love most generally. Sometimes, however, those who love most deeply are not man and woman who can celebrate their union in marriage, but two people of the same sex. What can be said to them, especially when we acknowledge that our deepest loves long for expression?

First of all, we must affirm that the strongest, most committed loving friendships should always be encouraged. They are creative of Christian life. The difficulty for the Church has always come when those friendships are homosexual and seek to express themselves in intimate sexual activity. The Church has been unable to see how it can bless sexual relationships which lack absolutely a generative character. But some people argue that there should be no problem. They draw attention to the many sexual acts of fertile married couples which do not lead to conception, to the situation of those sadly incapable of having children, and to those past child-bearing age. Their sexual love-making is not forbidden. Why should it be denied to homosexuals? That is the question. And they will often go on to affirm that homosexual and heterosexual loves have the same status. They are equals, alternatives. But the Church has a problem with that view. It seems to relegate the begetting, conceiving, and bearing of children to the margins of sexuality. And while sexuality must not be understood exclusively in terms of its generative purpose, that purpose cannot be ignored. There is, of course, much more to our sexuality than the capacity for reproduction, but that aspect remains nevertheless of its essence. So the Church has been unable to concede equal status to these loves or to see how it can bless those sexual relationships which do not merely lack this generative power in fact, but absolutely lack its character altogether. One further remark is appropriate. Some homosexual relationships will be truly loving. There may be sexual acts within them which express with delight a deep bond of love. They would put some marriages to shame. Why does the Church not bless them? We have met the reason already. For the Church, sexual intercourse does not depend only on the interior disposition of the individuals, however excellent that may be; it is a creative symbol and expression of love reserved for that deepest and most absorbing kind of loving union which bears fruit in new life. That is something of which homosexual love is incapable.

To those committed with care and generosity in a homosexual love, such comments may well seem very bleak. Homosexuals have often felt misunderstood by the Church and rejected by it. It is an indictment of the Church's pastoral care when that feeling is justified. And the same feeling has frequently been shared by those married couples who have agonized over the dilemma posed by the Church's teaching on contraception. Let me conclude this part with three observations.

First, what the Church teaches is not final. I do not mean by that that its doctrine is loosely provisional and of no consequence. But the final word is never uttered in this life. In these matters of Christian living, when the heart is pure and in search of love, fresh perspectives will open up. We must always be full of hope. As our capacity for love matures and we grow in wisdom, we must be confident that what now may seem such grave obstacles to the fulfilment in love which we desire, will be overcome. Love is our destiny.

Secondly, the teaching of the Church on these matters has regularly been the object of severe secular criticism. If our own feelings about it also are very negative, it would be good for us to examine our viewpoint with careful honesty. We should ask ourselves whether we have not been influenced too easily by society's standards. If we turn on them the critical appraisal we normally reserve for the Church's view, we may surprise ourselves. After all, the sexual attitudes of western society have not proved to be a panacea. A more detached scrutiny may reveal a standpoint with which we would not wish to be associated too closely.

And thirdly, in these matters very easily we may find ourselves in a part of life's battlefield where we suffer defeat. One insight missing in our society, has become familiar in these pages. We are called to share in the life of Christ. As his disciples, faithful in love, we will find ourselves in conflict at times with our world which is so scarred by sin. Once again, it is not a question of immediately identifying any hardship as our cross. We must struggle to overcome our handicaps. But if we are faithful in this sinful world, conflicts will arise, there will be a wounding. And the wound is most painful and bewildering when it is inflicted as we try to be generous in love. Those are the circumstances in which all of us, married or single, heterosexual or homosexual, may most easily suffer intensely. On reflection we realize that those were the circumstances too in which Christ was crucified.

(vi)

I have concentrated on marriage in these reflections on love, because in that relationship most usually are we able to earth in our daily lives Christ's command to love. Some of you may, however, have come to the conclusion reached once by a couple I was seeing in preparation for their wedding. They felt that I had made out so strong a case for married love as our means for fulfilling Christ's commandment, that they commented, 'But you're celibate. Where does all this leave you?' It would, of course, be a mistake to imagine that those who are unmarried, whether by chance, or choice, or by celibacy, are incapable of love, but we need to look at the character of their love.

In the first place, no Christian can manage to love God without loving other people as well. We know that. In the words of the first Letter of John, 'he who does not love his brother whom he has seen, cannot love God whom he has not seen' (1 John 4: 20). The love of neighbour is an integral part of our love for God. Moreover, as we have often stressed, Christian love is no mere abstraction; it is made real in particular relationships, with colleagues, acquaintances, friends, and in marriage. All the same, for the Christian these loves do not come to an end in themselves.

All love comes from God and finds its ultimate fulfilment in God. Human love indicates to us the way to that love. The more perfect it is, the more it reveals God's love. That woman's words to her husband in the London blitz, 'I'm not a whole person alone', may allow those who love to glimpse that their definitive wholeness will not even be found in such a pure love as that; that love itself is only a sign. Their definitive wholeness will be found only with Christ in God. As we mature in love, we will want to say with the great Saint Teresa, 'O my Lord, my Mercy and my Good! What more do I want in this life than to be so near Thee that there is no division between Thee and me?' (*Conceptions*, ch. IV).

Those who are not married, and particularly those who are unmarried by choice and are vowed to celibacy, have chosen a difficult path. They must not neglect friendship, but they leave on one side the powerful help which sacramental married love can provide. They seek out their wholeness with Christ in God more directly. They look to the union St Teresa celebrates. It may sound exhilarating. But no one should suppose that the scale of what is being attempted in this single loving is a simple matter. We are so made for human love that to be without its most profound realization can overwhelm us at times as the most tragic deprivation. Even those in vows, who have chosen to be celibate, will sometimes see their sacrifice as waste. Desolate, they may take shelter in an absorbing career or hobby, or simply slide into comfortable bachelordom. You will recognize by now the point we have reached. We are at the scene of another crucifixion. The response must be the same. Wounded in love, the unmarried must still remain faithful. Faithful in love, they will grow into that wholeness with Christ in God which is the destiny of us all. We cannot expect to follow Christ without suffering. What at one stage will seem like nothing but disaster, is transformed by faithful love into triumph.

The transformation is not immediate. No maturing process ever is. It is gradual. We must grow in love, grow into love. Its full glory can only come at the very end. As we mature in love, so do we look forward in hope to its fulfilment.

13

Hope and everlasting life

(i)

ON 2 October 1979 Pope John Paul II came to Harlem during his visit to the United States. There he told those who had gathered, 'We are an Easter people and "Alleluia" is our song.' The truth of those words should be plain to anyone who has followed these reflections. Jesus was utterly faithful in his love for his Father and for the entire human race. His interior life was moulded perfectly by that fidelity, so that his death on the cross and his rising from the dead are the supreme outward manifestation of that interior life. We are called to share in his life as individuals and as the Church. We have to be faithful as he is faithful. It is the principal unifying theme of this work and we have, on many occasions, encountered circumstances in which our fidelity may bring us defeat, but that fidelity sustained will also usher in new life. It is, as we have noted so frequently, the pattern of Calvary and Easter. Every sacrament is rooted in it. Christian life, stamped in faith and love, cannot avoid it. Death and resurrection mark us out. We are an Easter people.

And we believe that this idea is not just an uplifting way of speaking figuratively about times of trial when we have to cope generously. We believe rather that the Christ has died and risen, and that when we die and rise with him in this life, then he comes to dwell in us and we share in his life. We anticipate thereby a future state of glory, that is, our state of perfect union with God. That is the final state of salvation which is revealed to us in Christ. St Paul once called it 'the hope of glory' (Col. 1: 27). To be an Easter people is to be a people of hope.

(ii)

Hope is misty, elusive. It looks to the future. The Jews lived in the hope that the promises which the Lord had made to their forefathers would be fulfilled. In the Genesis account, as soon as Adam had sinned, God had given him the hope of salvation through his words to the serpent which have been understood as foretelling the birth of the saviour: 'I will put enmity between you and the woman, and between your seed and her seed; he shall bruise your head, and you shall bruise his heel' (Gen. 3: 15). When Abram was called, he was to become the father of a great nation; he was promised a new land and God's blessing (Gen. 12: 1–2). The deliverance of the Jews from their slavery in Egypt was to bring them to 'a good and broad land, a land flowing with milk and honey' (Exod. 3: 8). Much more could be added to illustrate this earlier understanding of the promise, expressed in terms of land gained.

With the prophets there is a shift of emphasis. It is no longer a matter of territorial gain alone, but of the qualities of the Kingdom, peace, salvation, light, healing, and redemption. The promise, however, is still understood in terms of worldly happiness. This conception is deepened further when it is seen as being the knowledge of God, brought about by a new law, planted within, written on the heart (see Ezek. 31: 33). But expectation is still restricted to this world.

Then there was a further development. People reflected upon the problem of evil. They realized that the innocent suffered as well as the wicked. The Book of Job wrestled with this difficulty. They realized too in the great fourth song of the suffering servant in Isaiah that such suffering could be redemptive (Isa. 53). By the time of the Book of Daniel, in the second century BC, a hope of resurrection had emerged: 'many of those who sleep in the dust of the earth shall awake, some to everlasting life, and some to shame and everlasting contempt' (12: 2).

That sketch is very brief, but it may be sufficient to indicate the general backcloth against which Jesus appeared, summoning the people to repentance and proclaiming that the Kingdom was close at hand: 'The time is fulfilled, and the Kingdom of God is at hand; repent, and believe in the gospel' (Mark 1: 15). This theme of the Kingdom is prominent in the gospels of Matthew, Mark, and Luke. Jesus preached about the Kingdom, told parables about it, and explained them to his disciples. Even so, they frequently failed to understand his message. There is that touching moment at the beginning of the Acts of the Apostles when Jesus is about to ascend into heaven. Still they ask him, 'Lord, will you at this time restore the Kingdom to Israel?' (Acts 1: 6). They had not yet learned the lesson which the Fourth Gospel shows Jesus teaching Pilate: 'my kingship is not from the world' (John 18: 36). We should not be surprised. We have seen already that the promises of God were deeply ingrained in the people as promises to be fulfilled in this world. Moreover, this tendency may explain something about Jesus' public ministry which may seem puzzling to us. He is seen little by little coaxing from his followers an awareness of who he is; then at once he forbids them to speak about it to others. So, for example, in the presence of the apostles he elicits from Peter the confession, 'You are the Christ'. Then, we are told, 'he charged them to tell no one about him' (Mark 8: 29–30). He seems to want to fulfil and frustrate this aspect of his mission simultaneously. However, this behaviour may be explained by Jesus' knowledge that those remarks about him would be misunderstood. There is an undertow of political tension in all the gospels. Positive steps were needed to avoid his being taken as a new king or political leader. These steps were not always successful. He was crucified on political charges.

Hope was to be directed towards the Kingdom that was to come. But matters are complicated, because Jesus also taught that the Kingdom was already present. When the Pharisees asked him about its coming, St Luke tells us that he warned them against looking for signs which might be observed, for, he concluded, 'the Kingdom of God is in the midst of you' (Luke 17: 20–1). The Kingdom is future, the Kingdom is present. Hope is misty.

Such tension, however, should not take us by surprise. The situation of a reality achieved, yet still waiting to be achieved, is something we have met before when considering Christian morality: we have been reconciled; therefore, we must be reconciled. We have to become what we are. It is a constant theme of St Paul's teaching. We, he wrote, 'who have the first-fruits of the Spirit, groan inwardly as we wait for adoption as sons, the redemption of our bodies'. Then he added, 'For in this hope we were saved' (Rom. 8: 23–4).

Hope is an elusive virtue. It implies confidence and so a kind of certainty. Without such genuine confidence, there can be no hope. We describe that condition as despair. On the other hand, a confidence which cannot permit any chance of disappointment, is also misconceived. We call that state presumption. We find hope on the path between the two.

This hope, this sense of expectation that God's promises will be fulfilled, is an integral part of Christian life. Certainly, it looks to the future, to our perfect sharing with Christ in God. At the same time, it stirs in us an alertness to the present. Every aspect of this study has been trying to indicate the way in which we share in the Christ. His life, death, and resurrection have implications for us now, not only in the future; the Church is the people of God at the present time, not only in the time to come; the sacraments establish, build, and sustain the Christ-life in us now; that life bears fruit now in faith and love. We have to be on the watch for that divine presence. That watchfulness is hope made manifest. We look for the Christ who will come again, while we look for his presence now in our midst. Cardinal Newman once gave it as 'the very definition of a Christian—one who looks for Christ' (*Sermons on Subjects of the Day*, pp. 278–9). A Christian should be full of hope.

(iii)

When I was collecting my thoughts on this subject, I asked my colleague, Fina Bello, if she had any ideas on it. 'Well,' she replied, 'there isn't very much of it about.' It was easy to see what she meant. Britain has now such a high rate of unemployment that many people expect never to find work again, while many others have never been employed and no longer expect they ever will be. They view their situation as hopeless. Some areas of the world are controlled by totalitarian governments which deny people their basic human rights. Others are wracked by famine, flood or earthquake. Others again are battlegrounds which turn millions from their homes as refugees. Across the globe there are apparently insoluble conflicts, in Ireland, in the Middle East, in Africa, in Asia, in Central and Latin America. The list seems endless. Over us all there hangs the threat of nuclear holocaust. It is true, there is not much hope about.

The situation is aggravated by lack of faith. We noticed the point much earlier. As faith fades, death can seem all the more terrible. The promise of everlasting life has no meaning. It is replaced by 'the dread of perpetual extinction' (*Gaudium et Spes*, 8; p. 35 above). And it is possible to turn the screw still more tightly.

Amongst the Jews, we have seen, the problem of evil gave rise to a hope in everlasting life. Nowadays its influence is reversed. People wonder about a God who is said to be all-powerful, all-good, all-wise, and all-loving, yet who permits profound unhappiness, pain, and suffering. Some of this evil may be attributable to sinful humanity, but that cannot explain the natural disasters, nor does it supply an account of innocent suffering. These experiences can quench hope, but we must resist that conclusion.

There are three main points that we have to bear in mind. The first is very difficult in practice. We need to be aware of the part played by emotion and place it in perspective. Its power can prevent us from examining the situation more deeply. It can hinder our understanding. Suffering appals us so much that the emotion it generates can cancel every other consideration. But we are trying to look more carefully. Our emotional response, however valid, does not give us the complete answer.

Secondly, we are searching for a solution to a problem in *our* world. We need to beware the tendency that asks, 'Why could God not have made a world without famine, flood, or earthquake, and a world without disease?' But these evils are not sent; they are features of the kind of world we actually inhabit. God could no doubt have created the world that is described, but such a world would not be our world. Its existence would not solve our problem; it would avoid it. This tendency is a version on the grand scale of the trap into which critics sometimes fall, when they attack the book, film, play, art exhibition, whatever it may be, under review for not being a different film, book, play and so on. The objection is futile. We are what we are. We are seeking a solution to the problem of pain within this setting.

Thirdly, we have to ask, what is love? Consider two significant models. There is the loving parent who so cares for his children that he shields them from every possible discomfort. Immensely wealthy and powerful, he is able to create a cocoon within which his sons and daughters are protected completely from every hardship and difficulty. He is able to arrange matters so that their lives are never touched by pain. Suppose too that they are loving children, perfectly compliant. They never rebel. They accept with delight the perfect comfort with which they are provided. Indeed they are unaware of any other kind of life. Is this an idyll? Or have these people been fixed permanently in childhood? The second model is of a parent equally capable of arranging such a life for his children, but he does not. His love is perfectly supportive, but not perfectly protective. He does not wish evil on his children, but he wants them to grow up to be themselves which means that they must experience their world in all its ambiguity: sometimes a glorious world, sometimes tragic. Which parent do we regard the more highly? The first sees a life without hardship and pain as an absolute good; the second locates that good in maturity and love. Our answer must choose the second. But our criticism of the existence of evil implies that we wish God to be like the first parent, constantly intervening to shield us from the painful consequences of our human reality. Were he to do so, his love for us would be compromised. It is not compromised. He loves us without reserve and, in the person of Jesus, he did not even avoid those consequences. He came and suffered, died and rose. As we share his life, we rise in him. We are an Easter people.

Our hope is fulfilled here and now as we come to share in the life of Christ through our baptism. We have already been dead and now we have been raised in him. We have already broken through to new life. The sneer that our hope is all 'pie in the sky when you die', has no foundation in orthodox Christian teaching. Our hope is not only for the future; it is for the present time. That truth must be emphasized. Only by doing so are we able to discover the true basis for that future hope. We are not completely cut off from the future. What is to be will bring to fruition the life in Christ which has already begun. You may remember the words from the Preface at a requiem mass: 'the sadness of death gives way to the bright promise of immortality'. But it is the hope in that promise, the confidence in everlasting life, which so many people today find difficult, if not impossible.

(iv)

Let us go back. We should ask the question, what are we celebrating at Easter? Why is it a feast for us? We believe that Jesus of Nazareth was divine, the Word of God become flesh. But we do not believe that the Father was, so to speak, deprived of his Word. To speak in a way which reveals clearly the inadequacy of our language about God, we know that we have to say the Word came to dwell amongst us without ever leaving the Father. At Easter we are not celebrating a divine reunion. What are we celebrating? We believe that God raised Jesus from the dead. When we profess that belief, we are referring to the man, Jesus of Nazareth. That man was raised up. And he was raised. He did not return to life. He was not resuscitated. Nor was he a ghost, striking fear in those who saw him. He was raised. He passed through death to new life and those who have been baptized are brought to a share in that new life with him. In the light of this teaching it is easy to understand why St Paul should say that if Christ has not been raised, then our faith, that living bond which we believe we have with him, is in fact futile and we are still in our sins (1 Cor. 15: 17). But we believe that he was raised. That belief lies at the heart of the gospel; that is the good news.

The resurrection is not just wonderful for Jesus as a happy outcome to all his suffering, nor is it just a way of speaking about the way in which we can grow morally through the severe trials which may beset us. It is good news for us who share in his humanity, because it reveals to us our destiny. If we are faithful as he was faithful, we too will be raised from the dead to newness of life. Our hope is not only for this present life. We believe in everlasting life. At Easter we are celebrating not only because Jesus was raised from the dead, but also because the real significance of death has been revealed by what happened to him. Death is no longer to be feared as an exit to oblivion; it is the threshold to a new, transformed, glorified life with Christ in God. That is our belief.

Or is it? Few thoughts are likely to make us feel so insecure as our reflections upon life after death. We are puzzled by them, even afraid. We are suspicious in case our belief in an afterlife may be mere wish-fulfilment, a means of taming our terror of dying. We know that we can account for our origins and our life on earth without recourse to God. An explanation can be given. We ask ourselves why our death should involve God any more than our birth or life. We notice uneasily that it is not of a piece. We are aware also that other cultures and religious traditions speak of life after death. Their views in some instances may seem to us wildly implausible, in others perhaps persuasive. In this most insecure area of belief, however, we may still wonder why Christianity should be correct and the alternatives mistaken. Then again, we have a profound sense of our own personal unity and suspect that our bodily dissolution must mean our end. It is so easy to view the future with anxiety. Christians believe that there is life after death, but it is a belief which puts faith to the test.

I can offer no neat solutions to these difficulties; you would be right to mistrust me if I could. I offer some reflections instead.

Consider, first, our sense of unity. We believe in the resurrection of the body as well as life everlasting. Whatever that may mean in reality, it reflects our awareness of our wholeness: we are physical and spiritual. Our destiny cannot neglect either aspect. We may worry about what our ultimate life in God could be like, but it may be that the problem revolves around what we can imagine. It is good to remind ourselves from time to time that our imagining is not the limit of the possible. There have been generations of slaves who could never have imagined being free; it would not even have been a dream; they are free now. And sometimes we come across signs.

The twenty-sixth of March 1975 was the Wednesday of Holy Week. I spent that night in the flat of an old lady, Sarah McNamara, who was close to death. Her daughter, herself no longer young, had painful memories of being on her own the night her husband had died. She did not wish to repeat the experience with her mother and so I went to watch with her. By a sad chance she died while we both happened to be out of the room. But I shall not easily forget our laying out of that body. Sarah was more than ninety years old, her body a fragile wreck. I remember thinking that it would be difficult to imagine a corpse more expressive of mortality. Yet, as her daughter and I sponged her, and straightened her limbs, and tidied the bed, it was impossible to believe that Sarah McNamara had been annihilated. She had died physically. Her life had changed, but not ended. I regard this deathbed as a very privileged experience. It is true that I have never been afflicted with the terrible sense of finality when tending the dead which some people have known. I acknowledge the subjective element. All the same, the sense of life renewed was remarkable in the presence of so spent a body. It seemed to me to signal the life to come. I tell the story for the light it may shed.

Next there is the question of the views of other cultures and traditions. I can make no claim to much knowledge about them, but must observe that I am disinclined to allow them to nudge me into a state of indecision or indifferentism. Their interest in the afterlife does not reduce whatever might be said to the relative. It would be arrogant to assume that their views bear witness only to escapism from the stark reality of a dreaded oblivion. Whatever the truth may ultimately be, may these other beliefs not be an attempt to express what has been glimpsed as true about our destiny?

Finally, as Christian people, we have to bring to bear on our own fear of death the Easter experience of the apostles. When Jesus was arrested, they scattered. They were disillusioned, they lacked courage. Soon they came together again behind locked doors in the upper room. They were frightened. The Gospels make no effort to excuse them. It would have been easy to do. Then after an amazingly brief period of time they appeared in public again. They spoke to the crowds about Jesus, who, they said, had died, but had then risen from the dead. They were emphatic that they had seen him, and encouraged everybody else to share their new-found faith. The authorities were appalled. They arrested them, questioned them, flogged and imprisoned them. In due course they had some of them put to death. It made no difference. Even those scholars who are rather sceptical about the resurrection of Jesus will admit a difficulty here. The change in the apostles is too sudden and complete, too out of character with the men we have known, to be easily explicable. Something remarkable had evidently happened. Christians believe that they were restored and revitalized by their experience of the risen Jesus. At the heart of their conviction is a teaching about the true significance of human death and destiny.

At first, the fulfilment of the promise was expected very soon. The Lord's return in glory was thought to be imminent. When it did not take place and some who had been converted, died, there was bewilderment. So Paul wrote to the Thessalonians to reassure them: 'We would not have you ignorant, brethren, concerning those who are asleep, that you may not grieve as others do who have no hope. For since we believe that Jesus died and rose again, even so, through Jesus, God will bring with him those who have fallen asleep.' He went on to speak about the Lord's coming, about the dead in Christ who will rise first and then be joined by those who are still alive; 'and so', he concluded, 'we shall always be with the Lord. Therefore comfort one another with these words' (1 Thess. 4: 13–18). We should allow the apostles' witness to the resurrection of Jesus, made manifest in the life of the Church, to reinforce our faith.

Such considerations may ease some of our difficulties, but others remain. We wonder about the experience of death itself and about our existence in heaven or in hell.

(v)

The fear of dying is not the prerogative of the unbeliever. Death has been mentioned often in these pages. On many occasions the experience of defeat has been likened to dying, to our being crucified with Christ. Our physical dying may well be another such occasion.

Although some will have faced it years in advance and will meet the event itself with the utmost serenity, for many others it will be an anguished experience. They will dread the pain and the parting from those whom they love, while the fear of the unknown may also overwhelm them. We should not be ashamed of ourselves, if it seems like that to any of us. The joy of resurrection does not unmake the pain of dying. Easter transforms Calvary; it cannot make it never to have been. When someone has been widowed and they marry again happily, their new-found joy will not simply obliterate their past sorrow. And so here too, we may die in faith, but still feel keenly the wounding of our passing. What the other defeats in our life have foreshadowed and anticipated is at last made real when we come to die.

Difficult questions assail us. We have often noted that our condition is made up of a relationship between the physical and the spiritual. While we are conscious of the spiritual aspect of our being, our experience is rooted in the physical, the particular, and it is hard for us to bring home to ourselves what the transfiguration of our bodies might imply. We have difficulty imagining a condition, distinct from time and space, which is also bodily. Yet we believe that being in the body is essential to being human. To be separated from the bodily would be to become dehumanized.

Does death, therefore, cause our human dissolution? Once again, it is vital to remember the important truth that our imagination is not the limit of the possible. Like slaves who cannot imagine being free, we cannot imagine ourselves beyond space and time, but the limits of what we can imagine do not determine what may be the case. And to that truth we, as Christian believers, must add the New Testament evidence that Jesus rose from the dead, was seen, touched, and listened to by his disciples, as for a short while he walked and ate with them once again. At the same time, they did not always recognize him at once or he would appear in their midst, in spite of the doors being locked. You will remember that these accounts do not have to be read literally (pp. 31–3 above). They are rather meant to teach us about the reality of his rising, which was not a return from death, but a passing beyond it to a new and transformed life. His resurrection is the source of our hope.

You may ask whether I believe in everlasting life. In fact, by a coincidence the question was put to me with great seriousness earlier today (28 February 1985). I must answer that I do.

I can see clearly the problems with this position. I recognize that it is possible to give an account of human experience which makes death our final end. I acknowledge the mixture of motives and influences which have caused human societies to affirm a life after death. But no version of events can give satisfactory answers to every question. Were I to lose my faith, I would still have problems. My spirituality, my unlimited capacity for knowledge and love, would disturb me. As it is, I look into myself and perceive that spirituality. I cannot imagine how my physical nature will remain in union with the spiritual after my death, but I see that as a weakness of my imagination, not as determining what is really possible. I know that I am one. And I believe what the New Testament teaches: that Jesus rose from the dead and revealed his transfigured humanity to his disciples. That belief is confirmed for me by the change that overcame them. Cowardly, untrue followers become faithful friends who will accept every hardship, even death, in order to make him known and bring others to share his risen glory. Of course, there is no conclusive proof. It is a matter for faith. And I believe in the resurrection of the body and life everlasting.

(vi)

When we speak of death as Christians, we do not refer simply to the physical act of dying, now commonly associated with brain death. We are talking about our total handing over of ourselves to God. The decisive death is the death of Jesus. At our baptism we were baptized into his death, so that we might walk with him in newness of life. The familiar theme returns. Each sacrament in its appropriate way draws us into his dying and rising. Through our share in that reality we are members of the Church, God's faithful people, faithful in love as Jesus was faithful, faithful in defeat and triumph, in death and resurrection. The interior disposition of fidelity is fundamental and it is perfected day by day as we accept its consequences in all the events of our lives. All Christian life, therefore, is sealed with death and resurrection. Every day we try to perfect this handing over of ourselves to God in perfect love. Our physical death is the final climactic expression of that interior disposition.

And so we may say that all Christian dying is martyrdom. We associate martyrdom most readily with a bloody and painful death. We call martyrs those whose deaths display most vividly their total surrender of themselves to God's will in perfectly faithful love. Superficially, a martyr, welcoming death, may look like a suicide. But those who commit suicide, can no longer see life as an absolute good and so are trying to escape from it. The martyr too can see that life is not an absolute good, but he does not try to escape, until the time comes when to stay would be a betrayal of faithful love. Then he accepts his death. St Ignatius of Antioch showed in his letters his care for the Church, but he had also recognized in his arrest and execution a supreme opportunity to display his faithful love. In words which have often been misunderstood, he wrote to the Church in Rome, where he would in fact be put to death: 'I am truly in earnest about dying for God' and he urged them not to put obstacles in his way. Shortly afterwards he continued, 'I am yearning for death with all the passion of a lover. Earthly longings have been crucified; in me there is left no spark of desire for mundane things, but only a murmur of living water that whispers within me, "Come to the Father" ' (Ignatius, *Romans*, 4, 7). The extreme example may illustrate the point that faithful love is our supreme good. What is displayed so dramatically in the death of a martyr is a disposition to be perfected in each one of us until the moment we die. By the time of our death we wish to have perfected our fidelity in love. For us death is definitive; at death our destiny is decided.

(vii)

There follows judgement. Christians speak of a personal and a final judgement. We should not allow these terms to baffle us unnecessarily. We can speak only in the terms that we know and we are heavily conditioned by our experience in space and time. What this way of speaking reflects is a basic perception about ourselves and our condition. We are individuals and we are members of a community. We have responsibilities on both accounts. They are not in conflict. They affirm each other mutually.

We have stressed so much our particularity and now we can see it bearing fruit. We are called to love and our loving becomes real in our particular relationships. And we have seen also that those individual loves are not ultimately exclusive; they make possible our fulfilment of the great commandment of love. We can now add that the mark of the authentic community, one grounded in love, is a community within which particular loves will flourish. The *truest* living for self is not selfishness, but a living for others; the truest service of others perfects the self. We believe that we will be judged according to our love and service.

God is our judge. He reveals our true condition. His judgement is not imposed, but it unveils the identity which corresponds to our self-knowledge. Our destiny will be the corollary of our state. We do not believe that he either keeps us in heaven or sends us to hell. If the deep decision which has given our life its significance is a radical rejection of love, deliberate infidelity, then his presence will be intolerable to us. Those who have banished all love from their lives will not be able to withstand the glory of the divine presence. And here it will be instructive to refer to something explained most helpfully by Fr. Herbert McCabe.

In general people work with two pictures of hell, the lake of burning sulphur and a place where death is not accepted, a place, that is, to use the terms familiar to these pages, where people are so caught up in their own defeats, so lacking in generosity and love, that they have never moved through from defeat to triumph, from death to resurrection; so it is a place where God is absent. The *first* picture, the more traditional one, is concerned with the character of human beings. Accordingly, it presents an image of a hell, which we would certainly want to avoid. We have no wish to be scalded by a boiling kettle, so the lake of burning sulphur is certainly repulsive. The second, the more modern picture, is concerned with the character of God. It teaches us that we go to hell of our own volition by resisting God and his love. And problems arise when we confuse the two. Mary McCarthy in her *Memories of a Catholic Girlhood* gave a clear example of the consequences of this misunderstanding. She had been taught to see God as sending people to hell and naturally she rebelled against this monster. 'I do not mind if I lose my soul for all eternity. If the kind of God exists who would damn me for not working out a deal with him, then that is unfortunate. I should not care to spend eternity in the company of such a person' (*Memories of a Catholic Girlhood* (London, 1957), p.xxxiv). It would be easy to sympathize, were this the case. And others conclude that, if hell is really the absence of God, they could handle it: they have not been concerned with him during this life; they can manage without him for eternity. But God does not send us to hell, as Mary McCarthy supposed; we send ourselves there. The *second* picture explains how we go to hell: we isolate ourselves from him. And the first picture comes in to correct any notion we might have that such isolation is tolerable. McCabe concludes: 'The fire of hell is God. God is terrible and no man can look upon him and live, he is a consuming fire. To be safe in the presence of God you must be yourself sacred, you must share in God's power and life. To have come into the presence of God without this protection is damnation' (*The New Creation* (London, 1965), p. 212). That is the basic biblical image of hell and should be complemented by the other: hell as the inability to accept death and so to die in Christ. Without that death there can be no resurrection. Those who will not accept death will remain for ever exiles from Easter.

But heaven is eternal Easter. When we have been faithful in love, that love at the last will be perfected in us. Some people contemplate the possibility with mixed feelings. Heaven going on and on can seem so boring. But it does not go on and on. Remember your happiest experiences. Some indeed, particularly in childhood, like a holiday or a visit to the pantomime, seem to have passed in a flash, but others, particularly as we mature, affect us quite differently. We look back on them, astonished to realize how brief they were, for they had an intensity which left an impression of something which lasted and lasted. In such circumstances we sometimes say that time stood still. Is it impossible to imagine a fulfilment in love so perfect that it is caught in the eternal now? Of course, it is; but we can try. That would be heaven.

I must mention one further subject before I conclude. When we look at ourselves realistically, we know (I hope) that we have not rejected love, but we also realize that the quality of our fidelity is often poor. Our decision to be faithful in love has not yet been made perfect in us. There is still an element of that deathly clinging to the self which inhibits our death in Christ. The Church has perceived a gap between our state at death and the perfect state we must achieve for heaven. We need to be purified and that process is expressed in the doctrine of purgatory. It is not a second chance, a second death; rather we work out the consequences of our decision for Christ, which have previously been impeded, but not annulled, by our sins. If the example may be allowed, we will be like people who have arrived at some occasion full of enthusiasm. Only on our arrival, indeed only by arriving, do we see ourselves as we really are, so ill-prepared for what is about to take place. We beg to be excused to put ourselves in order. But for those who have died in Christ perfectly, for the martyrs, there is no purgatory.

(viii)

We had seen earlier that each of the sacraments draws us into the death and the resurrection of Jesus. Now we have found that each of these great virtues sets on us the seal of his dying and rising. Faith is not a shelter from defeat, but a way of suffering defeat and passing on from it to victorious new life. Love is a battlefield in our sin-scarred world, where deep wounds are inflicted; when they are accepted in love, we rise again. And hope is that virtue which both seeks out the presence of the risen Christ in our midst day by day and directs our gaze on from death to our perfect risen life with Christ in God.

Let us turn next to Mary, the Mother of Jesus. She has often been presented as the perfect disciple. She should bring these virtues into sharper focus.

14

Mary, the mother of Jesus

(i)

MY work as Catholic Chaplain at Oxford University is pastoral. However, as I had done some research on Cardinal Newman, I was invited to join the Theology Faculty. I accepted and have some extremely light teaching duties. In particular, each autumn I take an undergraduate group with Dr Geoffrey Rowell, the Chaplain of Keble, and we examine some Newman texts. Newman, you may know, has been influential in bringing the Church to accept that doctrine develops. He discussed the idea formally first of all in a sermon preached in 1843. It is one of his writings discussed each year by that student group. I mention it here for a particular purpose.

Newman chose as his text a verse from St Luke's Gospel: 'But Mary kept all these things, and pondered them in her heart' (Luke 2: 19). One year the man who introduced our discussion was very appreciative of the sermon, but critical of the text. He objected to Newman's introductory remarks on Mary as intrusive. References to her, he argued, obscured the Gospel message. I do not pretend to remember his comments clearly, but I recall that they triggered off in me a vivid sense that this position, which is after all common enough, was very curious.

We believe that Christian faith is based on Jesus of Nazareth as true God and true man. As he was truly a man and as the truth of the humanity is taken to imply the highest degree of moral perfection (to use those terms in this context), it is bewildering to combine belief in that perfect humanity with indifference to his mother. If Jesus was perfect, he will have loved his mother as much as anyone. Those who wish to follow him, will want to love as he loved. That is not an excuse for every lunatic excess. Genuine loving is not fanatical. It unveils what is true, embracing it warmly and so protecting it from the chill of narrow calculation. Mary does not detract from Jesus. Our care for her flows from our love for him. So what account of her part in our beliefs can be given?

(ii)

The birth of Jesus is the creative source of Christian belief and life. Our religion rests on an understanding of the relationship between the divine and the human which is revealed in him. In him the divine and the human are perfectly one. Building on that, it has been a constant refrain of these thoughts that the reality of divine presence is not signalled to us as something additional. Although the divine and the human are never to be confused, the divine which is distinct, is perceived and known in and through the human. Whether we have been considering the Church or the Scriptures, dogma or papacy or sacramental life, the pattern has been the same. Revealed religion is not blueprint religion. What has been given by God is made known in man. And supremely in the man Jesus, God was recognized as present amongst his people.

'When the time had fully come,' Paul told the Galatians, 'God sent forth his Son, born of a woman' (Gal. 4: 4). We see in the accounts of the conception and birth of Jesus that respect for the human in the exercise of divine action which we should by now expect. God is not acting without consideration for the mother of the child. The angel Gabriel was sent to Mary of Nazareth. She was told of her favour with God and of the child she would conceive and bear by the power of the Most High through the indwelling of the Holy Spirit. And Mary said, 'Behold, I am the handmaid of the Lord; let it be to me according to your word' (see Luke 1: 26–38). Mary was indeed informed of her role, but the Church has always understood her response to have been decisive. It has echoed the words of Elizabeth: 'blessed is she who believed that there would be a fulfilment of what was spoken to her from the Lord' (Luke 1: 45). Her readiness to co-operate with the will of God was a necessary condition of that fulfilment. Her words of acceptance stand as the motto of Christian faith. It does not, therefore, surprise us to find that, when a woman in the crowd once called out to Jesus, 'Blessed is the womb that bore you, and the breasts that you sucked!', he should have pointed to a more profound cause of blessedness. 'Blessed rather', he answered, 'are those who hear the word of God and keep it!' (Luke 11: 27–8).

Her faith was, moreover, creative. The interior disposition which enabled her to respond to God's invitation, was united with her acceptance of the consequences of that disposition. Her acceptance of

God's word allowed her to conceive the Christ-child. As by her faith she was the one in whom the Christ-child dwelt, so in her faith is she a model for the life of the Church, in which Christ dwells. As by her faith she gave birth to the Christ, so in her faith is she the model of every Christian. Strange as it may seem to say it, all Christians, men as well as women, are called to be mothers of the Christ. Their lives in Christ, founded on faith, will make his presence real within them and they must bring him forth for the salvation of the world. People sometimes talk of our finding Christ in others; the burning issue is whether they will find him in us. The true disciple will always be like Mary. We are meant to be like her, not as an alternative to being like Christ, but because our likeness to her is a sure guarantee of that very likeness to the Christ which makes us his disciples.

And Mary's faith illustrates ours in another way. She was the beloved Mother of God, highly favoured, full of grace, but these immense privileges did not guarantee for her a life of ease. Living faith, we have seen, is not just a source of comfort and delight. Those who have faith share the burdens of a world marked by sin, in conflict with love. And so we find that Simeon warned her about the sword that would pierce through her soul (see Luke 2: 35), we read of her lack of understanding when Jesus remained on in the temple at Jerusalem after the feast of Passover, when he was twelve (Luke 2: 41–51), we remember that it is possible that she was amongst those relatives who came to take him away, fearing that he was out of his mind (Mark 3: 21), and we recall that on Calvary she took her place beside the cross (John 19: 26–7). For a long time it has seemed to me that the sight of his mother in the crowd must have been one of the most testing moments of Jesus' passion. Even on a view that argues for Jesus having a very limited awareness of why his service of the Father should have brought him to this end, it is possible to say that he had comfort in the knowledge that he was obeying his Father's will. He had remained faithful. Then, when he saw his mother, he would have realized that his own suffering did not in fact complete everything: his passion was the cause of sorrow in her. His awareness of defeat must have been intense. So there is plenty of scriptural evidence that her faith was put to the test.

Mary comes before us in the Gospels as a model for our faith, responding to the Father, obedient to his will, and wounded in love. But you may have felt a difficulty. I have spoken of her part in our redemption as evidence of God's respect for the human condition.

But does that respect seem shallow in the light of the Church's
teaching on her virginity?

(iii)

The Church teaches that Mary was a virgin before the birth of Jesus,
at his birth, and after his birth. The question of a virgin birth, that is
to say, a painless birth in which the hymen was not ruptured, is not
our concern; whatever may actually have happened, it cannot be
affirmed that the biological manner of Jesus' birth is an object of our
faith. Nor is there space here to review those scriptural references to
Jesus' brothers and come to a decisive conclusion. It is a matter of
history and beyond our scope. The crucial issue concerns the virginal
conception, the doctrine, namely, that Jesus was conceived without
the intervention of a human father. That is the question which has
been most controversial. However, the first point to be made is to
note the relative novelty of the controversy. From 200 to 1800, the
belief of the Church was virtually unanimous. The arguments have
arisen over the last two hundred years. And a second point is to
remember not to force the Gospels to say things which they never
had in view. These documents were written to teach people about
Jesus of Nazareth by proclaiming to them the significance of his
birth, life, death, and resurrection. Significance is primary, not the
story. Of course, the significance is communicated by the story, but
attention is not concentrated upon the historical detail of what took
place; the evangelists were concerned rather to present its underlying
meaning.

Guided by that perspective we must notice that both Matthew and
Luke affirm plainly the virginal conception of Jesus. According to
Matthew, when Mary 'had been betrothed to Joseph, before they
came together she was found to be with child of the Holy Spirit'.
Joseph resolved to send her away quietly, but had a dream in which
an angel assured him that 'that which is conceived in her is of the
Holy Spirit' (Matt. 1: 18–20). In the Gospel of Luke, the angel
Gabriel tells Mary, 'You will conceive in your womb and bear a son
and you shall call his name Jesus'. Mary asks, 'How can this be, since
I have no husband?' And the angel replies, 'The Holy Spirit will
come upon you, and the power of the Most High will overshadow
you; therefore the child to be born will be called holy, the Son of
God' (Luke 1: 31, 34–5). This teaching is clear. It is also unique. The

parallel examples which are often proposed are drawn from pagan world religions and from Judaism. In no case, however, is the parallel genuine; all presuppose some form of intercourse. But Mary is not impregnated by a male deity or element; her child is begotten through the creative power of the Holy Spirit. What then is the significance of this teaching?

Various possibilities can be suggested. Some will see in the virginal conception a safeguard for the divinity of Christ. During the centuries when this doctrine was maintained virtually unanimously, it is true that those who opposed it usually did so because they denied as well that Jesus was divine. Virginal conception and divinity have gone together. However, it is impossible to maintain that belief in Jesus' divinity *requires* the virginal conception. Jesus would still be the divine Son of God if he had been conceived and born in the normal human manner. Indeed, that is what a theology based on incarnation would, if anything, lead us to expect.

If it is not a safeguard for the divinity, some may ask, is it evidence of a lack of belief in the humanity of Jesus? But that position too cannot be maintained. There is nothing in the New Testament nor in the witness of the early Church to suggest that Jesus' virginal conception was seen to compromise his humanity. Thus, to use one example, we find Ignatius in his letter to the Church at Smyrna, combatting Docetism, the heresy which argued that Jesus never bore 'a real human body'. He set before them their firmest convictions about Jesus: that he was 'truly of David's line in his manhood, yet Son of God by the Divine will and power; truly born of a Virgin; baptised by John for His fulfilling of all righteousness; and in the days of Pontius Pilate and Herod the Tetrarch truly pierced by nails in His human flesh (a Fruit imparting life to us from His most blessed Passion), so that by His resurrection He might set up a beacon for all time to call together His saints and believers, whether Jews or Gentiles, in the one body of His Church' (Ignatius, *Smyrnaeans* 1). Here evidently Ignatius, writing about 117, saw the virginal conception of Jesus as being of a piece with such events as his baptism by John and his crucifixion. Moreover, as it is something he used confidently to refute Docetism, he obviously did not regard it as compromising the humanity of Christ.

A further possibility remains. Does the doctrine bear witness to that negative attitude to sexuality which has come to be associated so readily with Church teaching? Once again, it will not do. The view which looked upon sexual intercourse as defiling arose much later. Those who adhered to it may in fact have read their prejudice back into this doctrine, but its origins cannot be found in that way. At the time the Gospels were written, the contrary view prevailed. Sexual relations were valued as good and wholesome. It would be a mistake to account for this doctrine in terms of a later pessimism.

I suggest a different approach. We should remember that Christianity is a religion of salvation. Its origins are to be found in the calamitous estrangement between God and man which we express in the doctrine of original sin. Essential to our understanding of what that sin implies is a damage so terrible that it cannot right itself, nor can the human race make matters right; that sin has disabled them. Salvation can come only from God. And we believe that God intervened. He has offered us that salvation by sending us his Son who became a man for our sake. He was fully and completely a man, truly a man. But he was also God. The divine and the human were united in him perfectly without being confused. Through that perfect relationship of humanity to divinity in him, our relationship with God can be restored. But that restoration depended upon God's intervening.

It is necessary to proceed carefully. Divine intervention did not require a virginal conception. As we have noted already, Jesus would have been nonetheless the divine Son if he had been conceived in the normal manner. So what is the significance of a virginal conception which, on the one hand, is not needed to safeguard Jesus' divinity and, on the other, cannot be attributed to a flawed view of his humanity or a negative attitude towards sexuality? Only one really plausible explanation remains. This doctrine was taught because it was true. How else can we account for something that is both a unique teaching and one not required of necessity? Our dogma, we believe, tell us of God's actions. It may be helpful to compare this doctrine with the resurrection. When we spoke of that, we said that God raised Jesus from the dead. We did not identify that belief as a conviction about the empty tomb, but we affirmed that the tomb was empty, and located its significance at a secondary level: the fact of the empty tomb earthed the spiritual dimension of God's act of raising Jesus in the physical. And here, when we speak of the virginal conception, we say that God sent his Son as a man. We do not identify that belief as a conviction about Mary's biological condition, but we affirm that she conceived the child and remained a virgin, and locate the significance of her virginity at a secondary level: her actual virginity earths for us the birth of Jesus as God's intervention.

The doctrine of Mary's virginity does not compromise God's respect for the human condition. However, a related doubt may nag at us. We have recalled that Mary's acceptance of the angel's message was decisive, but we may wonder whether her response was free. We must consider the mystery of her love for God and God's love for her. We must turn to the doctrine of the immaculate conception.

(iv)

The doctrine of the immaculate conception is one often held by non-Catholics and unbelievers. They may not recognize it, but that is generally the case. The point was well illustrated on the occasion when Cardinal Heenan (long before he was made a cardinal) was trying to explain this teaching to a hostile interviewer on television. He was having no success, so he tried a new approach. 'Tell me,' he asked, 'do you believe in original sin?' The man dismissed the notion with a laugh. 'Then', Heenan explained, 'you believe in the immaculate conception.'

Difficulty arises most frequently because people confuse it with the teaching about the virginal conception of Jesus which we have just been considering; that it should be designated 'immaculate' is seen as further proof of the Church's lamentable attitude to sexuality. That view, of course, is mistaken. The immaculate conception of Mary refers to her conception in the womb of her mother, a conception which took place in the normal sexual manner. It is called immaculate to express the belief that she was preserved from sin from the first moment of her existence. We must begin with her freedom from original sin.

One of the earliest ways of understanding our redemption saw it as a re-enactment of the scene in the Garden of Eden, that terrible meeting between the serpent, the woman, and the man. The characters remained the same, but the play had been largely re-cast, for there was now a new Adam and a new Eve, and the new Eve was the mother of the new Adam. This way of looking at Mary as the second Eve goes back to the second century. It led easily to a very influential conclusion: if Mary was the second Eve, then she could not have been inferior to the first. So as the original Eve had been born without sin, how could Mary not have been sinless also? Here we find the basis for the doctrine of the immaculate conception, but it raises two obvious questions.

First of all, if Mary is without sin, then it may be asked whether she herself needs to be redeemed. The rest of us plainly need a redeemer. We know we are sinners. But why should she have to be redeemed when she is already sinless? The question is a good one, because it forces us to look more carefully at the nature of her sinlessness.

Some people think that original sin so contaminated mankind that anyone untouched by it would be quite distinct from common humanity. In this view, to sin has become an essential part of being human. But Catholic teaching is different. We believe that the tragic wounding of our nature which we call original sin, although it enfeebled us, did not infect our very natures. Mankind is not *essentially* sinful. Mary's actual freedom from sin does not set her apart as a different kind of human being. It is a blessing on her, not grounds for her expulsion from ordinary humanity. And this blessing is a privilege. It helps to remember what a privilege implies.

We are not privileged when we receive our due. We are privileged when we are given something which otherwise we ought not to have or which at the very least we cannot expect. To declare that Mary is privileged in her immaculate conception, as Pope Pius IX did when he proclaimed this dogma, is to affirm that her human nature is exactly the same as ours. In other words, she ought to have been flawed, just as we are. By a privilege she was not. That privilege she owed to her motherhood of Jesus. She received it from him. He is her redeemer quite as much as he is ours. He preserved her from sin; he releases us. The preserving and the releasing are no more than two expressions of the same redeeming work. Mary immaculate was saved by her Son.

The second question is caused by the comparison of Mary with Eve. In the past, they were seen as two historical people, but no longer. And we may feel uneasy with a doctrine that rests upon a contrast between the legendary figure of Eve and the historical Mary of Nazareth.

The solution is found in a shift of perspective. The contrast never was primarily historical. It was always a contrast of type as well. Eve was the symbol of responsible co-operation in our fall; Mary symbolized responsible co-operation in our restoration. In today's world, so handicapped by literal-mindedness, we have to become more sensitive to the symbolic character of the part Mary played in our salvation. She is our pattern and example. However, we cannot allow perfectly proper talk about the symbolic to become a denial or evasion of history. We say that Mary was without sin, but the sceptical mind asks the obvious question: how can we know that? We possess no detailed biography.

Part of the problem with the doctrine is the negative and static language in which it is expressed. I warm to the idea that Mary's immaculate conception should be expressed today in positive and dynamic terms. We should think of a life dominated totally by love for God. Perhaps our own lives can suggest to us what that means. For myself certainly I can think of good friends with whom the bond of friendship has never been broken. There has never been a need for reconciliation. There may have been misunderstandings, bewilderment, pain, but the bond has never snapped and, I would say of such friendships, could never snap. I would hope my experience is not too rare. In any case I think that it is a life dominated by love for God that we meet in the Gospel tradition. Everybody knows that there is no snug text which states 'Mary was conceived immaculate', but every reference brings home to us the meaning of that title. We meet a woman caught up perfectly in love for God. Her response to the angel flows from that love. Like every loving response it is free and, at the same time, a foregone conclusion. A true love is compelling. Her life is overshadowed by the Spirit in a unique manner and she ponders upon what has happened to her in the birth of her Son. At times she is bewildered and unsure; at times, and evidently at the end, she shares deeply in his pain. Her faith is not a shelter from suffering. She is no plastic madonna. But her bond with her God is unshaken. Her life is dominated by her love for him and for the Son whom she bore. She takes her place by his cross and she joins the apostles when they reassemble later in the upper room. She has heard the word of God and believed it and been totally obedient to it. Her fidelity in love is triumphant. She never sins. That is what immaculate conception means. And once again, her example is significant for us.

We have not been preserved from sin as Mary was, but we share with her that call to holiness of which her freedom from sin was a part. And her privilege had a purpose: she was to be the mother of the Saviour. And our holiness has a purpose: for all Christians, as we have seen, must become in a sense mothers of the Christ. Love is creative as well as faith. Our lives must be so overwhelmed by our love for God that we bear the real presence of his Son within us and make him manifest in the world.

Mary is the model and pattern for our faith. She bears witness to a love for God that is without reserve. She can teach us about hope as well. We believe that she was assumed bodily into heaven.

(v)

The assumption of Mary is unintelligible if approached in isolation. As with the doctrine of the immaculate conception, we know that there is no neat scriptural peg which declares, 'And then Mary was assumed bodily into heaven'. But such pegs are rarely available and, even when they are, they are not always what they seem. You need only recall such examples as the Matthaean text, 'You are Peter and on this rock I will build my Church' in connection with the papacy, or the use of the marriage feast of Cana as the basis for the sacrament of marriage, to realize how complex the interpretation of texts can be. We have to probe more deeply.

Let us go back to the touchstone of Christianity, the perfect fidelity of Jesus in love which has won for us our salvation. We remember that that is not just something he has done. He did it for our sake. Through his indwelling Spirit we can share in his faithfulness. The Church is the community of all those who are faithful; the sacraments, God's acts of love for us, bring about our share in that faithful life; faith, hope, and charity are the distinguishing virtues of our fidelity. At every point, the death and the resurrection of Jesus are made present in our midst. But there is something else. This presence and the corresponding fidelity do not come into existence instantaneously. We saw, first of all, in our reflections on morality that this work of our redemption has been accomplished in fact: we have been reconciled; and that it yet awaits its accomplishment: therefore be reconciled. We saw also, when we considered everlasting life, that the Kingdom of God has been established already, but still looks to its perfect, eternal fulfilment. The tension is a consequence of our sinning. We have to grow to perfection in Christ. God does not impose himself, but he loves us and invites us to answer love with love. It is a gradual process.

If that is a fair summary of the state of the Christian as it has emerged in this study, we can now turn back to Mary. What have we discovered about her? We have acknowledged that she was the one who was privileged to be the mother of the redeemer. She did not deserve that honour. It was God's gift to her. And the Scriptures have indicated to us the kind of woman she was. We have noted her faith which made her accept without reservation what she understood to be God's will for her. She believed and she was obedient in faith. At the same time, living faith is not a hideaway and Mary's faith was not divorced from real life: it was tested, she was at times bewildered, she was wounded by it; it need have been nonetheless perfect for that; it was drawing her into a share in the death and resurrection of her Son. We have noted also the love of God which utterly dominated her life. We believe that hers was a life overwhelmed by love and so untouched by sin. Again, she did not deserve that privilege. It was a part of God's redemptive gift.

We must draw these two thoughts together. If to be saved is to share perfectly in the faithful life of the crucified and risen Christ and we find in Mary a life given over to faith and love without reserve and sealed with his death and resurrection, then, we may ask, what is her destiny? We have no evidence that her response to God was qualified, that her death in Christ was imperfect. There is nothing to suggest that there was in her the tension which is the consequence of sin. So what can separate her from her God once her life on earth has ended? Only one answer seems possible: nothing at all. At her death Mary was assumed into that state of union with God which was consonant with her privileges and the perfection of her life of faith and love. At the moment of her death, she was perfectly redeemed. The Church declared her assumption to have been bodily to underline that perfection, for bodiliness, as we know, is essential to complete humanity. And as her life was perfect in faith and love, so her destiny can show us our grounds for hope. Or can it?

In 1979 Rome warned against interpretations of the destiny of those who died in Christ which deprived Mary of her uniqueness. There should be no question of that. Her privileged position as the sinless mother of Jesus and the perfect disciple is unassailable. All the same, such a privilege does not withdraw her from her place as our model and guide. In her destiny there can be found the pledge of that union with God which is the abiding hope of all who live lives of faith and love in Christ. Mary displays for us that holiness which every Christian should desire and seek with steady confidence.

(vi)

Mary is the model disciple, perfect in faith and hope and love. She is the mother of Jesus and so the mother of God. She is the mother of all believers and so the mother of the Church. A priest friend of mine once remarked to me that he liked to think that when God made woman out of man in Genesis, the result had been rather disappointing, but when he made a man out of a woman, Jesus from Mary, the result had been perfection. Be that as it may, we have much for which to thank God in Mary. Reflection upon her role, her ministry, her life, can bring the invitation of Gospel life into such clear focus for us. Seeing the prize we can pursue it more readily.

This account of Catholicism is almost complete. We have considered in turn the Christ, the Church, the sacraments, and the virtues. One major task remains, to give an account of our belief in God as Father and Son and Spirit. It could not have been done properly any sooner. God has not revealed himself in a sudden blinding flash. We recognize him in the revelation which his Son provided, which was his Son. We must ponder on the revealed mysteries. And here too Mary can be our model. St Luke tells us twice that she kept the things which were done and said in her heart, and pondered on them (Luke 2: 19, 51). Once again, we must do the same.

15

God: Father and Son and Holy Spirit

Does it seem shocking to finish this account of Catholicism by considering God as Father, Son, and Spirit? I have had trouble placing it. When I began, I knew it was a subject that had to be treated and it seemed natural to start with it, or at least to examine it after presenting reflections on the Christ. But it would not do. They led most naturally to a study of the Church, which in turn pointed to the sacraments by which we are made its members, and then to the virtues which mark Christian life. Suddenly I had reached the end. Could the Trinity be relegated to an appendix? That was absurd. And then, all at once, the mist cleared. The end was the place where an account of the Trinity ought to be in this work. I have not been trying to supply some formidably exhaustive synthesis of Catholicism. As I mentioned at the start, my aim has been much more modest. I have wanted simply to help people who, for a variety of reasons, were looking for an introduction to Catholic faith and belief which would make some sense to them, which would allow them to get their bearings. Once that has been done, but only then, is it possible for them to probe the mysteries of faith more deeply. It was suddenly obvious that the mystery of God could be approached best in the light of perceptions about Christ and the Church, the sacraments and the virtues.

The Trinity is awesome. I once sent a friend who lives in Edinburgh the text of a sermon I had preached on the subject at his Oxford college, Worcester. He replied and thanked me, but confessed that what I had been saying had made little impression on him. Then, a year later, he wrote to me again out of the blue:

Sitting alone in the Dominicans' chapel after mass yesterday, I began, for no apparent reason, to think about the Trinity. In no time at all I realised that it is inconceivable that God the Father could exist without God the Son and that their relationship generates the Holy Spirit—indeed their relationship is the Holy Spirit—although I am still hazy about why the Holy Spirit has to be a person, but I can hardly expect to come to a full understanding of the Trinity in two minutes flat. What I do realise, however, is that instead of regarding it as a rather remote intellectual concept, the whole of my faith has to be grounded in the Trinity and the most important thing in my life is cultivating right relations with God and my neighbour.

Here is someone wrestling with profound truth to make it his own. We must do that. We believe in God who is three and one. Is that meaningless or is it mystery? We believe it is mystery, but we must still attempt to give an account of our belief, not least to ourselves. Too often Christian faith has affirmed that God is one and allowed a sense of the Trinity to fade. We have spoken of Jesus naturally, and indeed of the Spirit also at Pentecost. More recently the charismatic renewal has placed attention on the Spirit more firmly. But the conviction that our one God is three, has usually been weak. Let us try to do better.

(ii)

Our Jewish ancestors in the faith held to a firm belief that God is one. In the earliest times, however, they did not believe that there was only one God. They acknowledged the reality of the gods of the other peoples amongst whom they moved, but they despised them. Other gods may have existed, but no god was as powerful as their Almighty God, Yahweh Sabaoth, the Lord God of Hosts. And gradually, of course, they came to perceive what their belief implied. The Almighty God was not merely the most superior amongst the gods, the supreme God must be the only true God: there could be no other, but him.

We think immediately of the opening words of the ten commandments: 'I am the Lord your God, who brought you out of the land of Egypt, out of the house of bondage. You shall have no other gods before me' (Exod. 20: 2–3). We remember Elijah mocking the prophets of Baal at Mount Carmel when no fire came from heaven to consume the bull laid out for sacrifice: 'Cry aloud, for he is a god; either he is musing, or he has gone aside, or he is on a journey, or perhaps he is asleep and must be awakened' (1 Kgs. 18: 27). And what such texts imply was expressed vividly in the sixth century prophecy of Isaiah: 'I am the Lord, and there is no other . . . there is no other god besides me, a righteous God and a Saviour; there is none besides me' (Isa. 45: 18, 21). By that time, monotheism was firmly established amongst the Jews. It was adhered to without reserve.

This understanding of God, however, was also rich and subtle. Think back to the account of creation at the beginning of the Book of Genesis: God creates; his Spirit moves over the face of the waters; he creates by speaking, by his word: 'and God said, "Let there be light"; and there was light' (Gen. 1: 1-3). Wisdom was presented as an eternal attribute of God. So, for example, wisdom speaks: 'The Lord created me at the beginning of his work' (see Prov. 8: 22-31), where the sense of the whole passage indicates that the beginning is not the first, and so a part, of a series, but the origin of all things in God, an eternal attribute of the eternal God. The Spirit was shown as the breath of God: it made the dry bones live (Ezek. 37). It is so difficult to be exact about the spirit. We think of natural phenomena like wind and fire which make things come alive: but spirit is not identified so easily. It should go without saying that neither wisdom nor spirit were perceived as personal. There is no trinitarian doctrine in the Old Testament.

My intention in making those brief comments is very modest indeed. The doctrine of God in the Old Testament calls for vast learning and close investigation. My purpose is limited to indicating that the deepening perception that God is one did not prevent sensitivity to the complexity of the divine mystery.

(iii)

The Christian tradition is rooted in Jewish monotheism. There can be no question of that. It is one of the factors which makes the development of trinitarian doctrine fascinating. In the circumstances the last conclusion we should expect would be the identification of the Christ as God and the Spirit as God. And yet, little by little, cautiously and with various false moves, that was what was done. In fact, very early in New Testament times first a diadic and then a triadic pattern of the Godhead begins to emerge.

Jesus is Son. His intimacy with the Father is a constant theme of the Gospels. The controversial title, 'Son of Man', is rich in associations. It is used frequently in the Gospels and yet rarely elsewhere, which may suggest that it is preserved on account of the use which

Jesus himself made of it. And we can recall the famous words: 'All things have been delivered to me by my Father; and no one knows the Son except the Father, and no one knows the Father except the Son and any one to whom the Son chooses to reveal him' (Matt. 11: 27; Luke 10: 22). This sense of the communion of Father and Son is then heightened further in those passages which include the Holy Spirit. Read once again St Mark's account of the baptism of Jesus:

In those days Jesus came from Nazareth of Galilee and was baptized by John in the Jordan. And when he came up out of the water, immediately he saw the heavens opened and the Spirit descending upon him like a dove; and a voice came from heaven, 'Thou art my beloved Son; with thee I am well pleased.' (Mark 1: 9-11)

And what is expressed by Matthew, Mark, and Luke is presented in a more fully developed form in the Gospel of John. Jesus is identified as the Word of God made flesh. He speaks of God as his Father in the most intimate way. He teaches that no one can come to him unless drawn by the Father (John 6: 44). He and the Father are one (John 10: 30). And the Spirit is to be given when he is glorified (John 7: 39). When the Spirit comes, he will teach the disciples all things and bring back to their minds everything that Jesus had taught them (John 14: 26).

The perception of God as Father and Son and Spirit is very clear in the New Testament. It is not a proof of trinitarian doctrine, but it bears witness to the experience of the early church. In the letters of Paul, in the Acts of the Apostles, and in the other New Testament writings we have evidence of members of the early Christian community attempting to give an account of their experience of God and the faith that had come to birth in them. They were over-whelmed by their sense of God as Father. As they asked themselves about Jesus, they found themselves unable to avoid the conclusion that he was divine, as well as truly a man. And they pondered deeply on their experience of the Spirit which came upon them and, as they believed, made them new. It is a cause for wonder that Paul should close a letter to the Corinthians, written probably in 57, less than thirty years after the death of Jesus, with the words: 'The grace of the Lord Jesus Christ and the love of God and the fellowship of the Holy Spirit be with you all' (2 Cor. 13: 14).

(iv)

Speaking generally, it is well known that it took a very long time for the implications of this recognition of God as three and one to be expressed systematically. Fierce controversies raged. There were those who so insisted on the oneness of God that the sense of God as three was reduced to a variety of ways in which the one God manifested himself: as law-giver in the Old Testament, as saviour in the New, and as sanctifier in our own day. That may seem plausible enough, but it is quite inadequate as an account of the inner life of God. And there were others who so emphasized that oneness that the Son and the Spirit were not truly divine at all: they were subordinate to the one God. It would be a mistake to become caught up with these controversies here. Let me restrict myself to two comments only.

First, what did the Church come to teach about God? It teaches that there is only one God. God is Father and Son and Spirit. Father and Son and Spirit are not three ingredients which mix together to make the one God. Each of the three is distinct from the others, has his own underlying reality: in other words, each is a person. But, as I have said, they are not mixed up, nor are they three parts of a whole. Each one is perceived as the whole God. They are united perfectly; they are distinct in what relates to them specifically as Persons: so what makes the Father specifically Father distinguishes him from the Son and the Spirit; what makes the Son specifically Son distinguishes him from the Father and the Spirit; what makes the Spirit specifically Spirit distinguishes him from the Father and the Son. Yet they are not three gods. There is only one God. Moreover, the perception of God as Father and Son and Spirit through the saving work of the Christ which revealed God as Father and brought down on us the Holy Spirit, while it shows us God in relationship with us, at the same time we believe, allows us a glimpse of the inner life of God as well.

Secondly, as I apologize for the abstract character of what I have just written, I would urge you not to try to grasp the mystery of the Trinity whole. I would commend the approach which I introduced before when we were considering the significance of Jesus as divine, as human, and as one, and studying the atonement in terms of the example of Jesus, the victory over sin and death, and the sacrifice of the Saviour. We took a cue from the appreciation of surrealism. So here. We can reflect that the Father is God, the Son is God, the Spirit is God, and God is one, each in turn, and allow those thoughts to make their impression upon us. It will not take place on the surface, permitting an easy description of the conclusion. It will be an exercise in that perceptiveness proper to living faith with which we should by now be familiar. What comes to be known part by part is made real within, as a vision of the heart.

To have said so much is not, however, enough. The experiences to which the Scriptures bear witness and the conclusions of the Church have too often seemed remote. I must try to help you grasp imaginatively what the Church believes about the mystery of God who is Father and Son and Spirit.

(v)

To state the obvious first: the problem of God as three and one is not mathematical. We are not trying to resolve how the same being can be a single being and yet, at the same time, three. We must reflect more carefully. We should consider some of our deepest concerns.

Some years ago I came across an article in *The Clergy Review* by a priest of the diocese of Arundel and Brighton, called Paul Edwards. He had entitled his piece, 'Today's World: Image of the Trinity' (*The Clergy Review* LXII, no. 6, (June 1977), pp. 242-5). I like it very much. It is a collection of ideas about the Trinity, gathered from looking around our world and society. He begins by emphasizing relationships. They form us, they make us who we are. And the Trinity differ only in their relationships as Father and Son and Spirit. He talks about the contemporary search for unity between the churches, between nations, between classes, within families, and notes that the unity we seek is one in which individuality, personality, are respected. He observes, 'We are trying to live as befits a Trinity, with perfect respect for each other's individuality and personality and yet in complete unity and love.' He speaks of the joy we find in sharing and of 'the sheer bliss of the Trinity who share everything perfectly between three'. He calls attention to the longing we feel to communicate to another what is deepest within us and remarks on 'the utter bliss of the three persons whose thoughts and feelings are so completely understood and shared, in the security of an utterly loving relationship'. He turns next to love and in particular the need to be properly and totally present in love, husbands and wives to each other, parents to children. The persons of the Trinity give each other that total and perfect presence which we desire. And finally, equality. He notices how we tend to work as superiors to inferiors, or to suffer from a sense of our own inferiority when in fact we are equal, though different. To combine a sense of our equality with acceptance of our differences is hard for us. When we succeed, we reflect the Trinity. In all these ways, and there are many others, the Trinity can teach us best about what we are seeking. Now let me concentrate most on relationship.

(vi)

Some time ago I attended a talk in which the speaker recalled a conversation he had had with a Benedictine monk. The monk was deploring how little consideration we give to the Trinity nowadays. 'You must contemplate and contemplate and contemplate,' he said, 'until your eyes bleed. And then perhaps you may be able to perceive some small truth.' 'And what small truth have you perceived?' he was asked. He replied, 'That relations are real.' Very simply I wish to explore that thought with you. Relations are real. It may seem banal, but sometimes a phrase becomes commonplace because it is profoundly true. Relations are real.

1. One of the delights of good friendship is the way it makes us grow. Subjects, hobbies, activities which previously had left us cold, or of which we had been unaware, are suddenly invested with an interest for us which they had never had before. New friends introduce us to fresh and unexplored fields. They expand us. But in the greatest loves, while this may be included, it is only secondary.

We have seen already in our reflections on love that to know and to love another most deeply is, first and foremost, to discover not increase, but a lack. Who we really are is revealed to us. Before the experience of that knowledge and love we may have thought ourselves whole, complete, autonomous, self-possessed. The shock of profound love is the recognition that that was not true. We realize that, before loving, all unawares, we had not been whole; that without the other we are not whole; without the other, unloved, we are diminished, cripples, handicapped. You may remember Jill Furse, writing to Laurence Whistler during the blitz: 'I'm not a whole person alone, and the edge of the tear hurts all the time.' Who we are, the wholeness of our being, is only made known to us when it is made real in the profound relationship of knowlege and love. And this self-discovery can disclose something even more over-whelming. For we believe that we are made in the image and likeness of God. Accordingly we can be confident that, when we recognize and know ourselves most truly, what we perceive can reveal to us, though partially, yet really, something true about God. It does so here. We believe in God who is Father and Son and Spirit. We do not believe that our God is somehow 'more God', because he is Father and Son and Spirit; we believe that who God really is, the whole, the one God, is realized in those relationships. They are not prejudicial to his oneness. Without them he would be diminished. He would not truly be the one God. Relations are real.

2. It is characteristic of good friendship to be able to endure separation without any great sense of loss, and on reunion to be able to pick up the threads 'just where we left off'. Close friends often remark on the ease with which they resume their companionship even after many years. That may also be true between lovers, but once again it is secondary.

To love deeply and to be unable to give expression to that love is a cause of intense suffering. Absence or loss is keenly felt. The most poignant example is bereavement. When the beloved dies, love does not cease. Precisely not. What makes the pain so intense is the continuance of love and the impossibility any longer of giving satisfying expression to it. And again, to give a very different example, the teacher who loves his subject and is utterly absorbed by it, will be consumed with the desire to share what he knows with others. The deepest knowledge and love cry out for expression. And we believe in God who is Father and Son and Spirit, not just one God, but the one God whose wholeness is realized and revealed to us in his relationships. We believe in a God who knows and who loves with such intensity that the knowledge and the love have to be expressed. The Son or Word gives expression to the Father's knowing, the Spirit to the love between them. Indeed, we say that in God the expression of knowledge and love is so perfect that it is indistinguishable from what it expresses, except in so far as it is the expression: Father, Son, and Spirit are distinct from each other only in their relationships. It is a profound insight which brings home to us the truth that the compulsion we feel for knowledge and love to be expressed, unveils for us something true about the God of know-ledge and love. Relations are real.

3. I have suggested to you that the deepest relationships reveal that we are complete, made whole, only with and through the beloved. But we do not thereby lose our autonomy. In more superficial relationships, of course, the bond may actually be a crutch, a support. That is something very different.

But when love is deepest and truest, although we would be diminished without it, nevertheless we are established by it in our own autonomy, integrity, independence. It makes each lover whole. It may not be obvious how this can be true when we only think about it, but the experience is clear, convincing, and unquestionable. The very power of the bond and the completeness that it reveals in the union, at the same time affirms the wholeness of each lover. And this too allows us to glimpse what we believe about God. For we believe that the one God is Father and Son and Spirit. We believe that this very union in relationship, the completeness of God which is made known to us in and through these relationships, does not compromise each Person, but on the contrary establishes their distinctiveness. We do not believe that the unity of God damages the integrity of Father and Son and Spirit. To each it is possible to pray, 'You *alone* are God, living and true'.

Relations are real. They reveal the true unity of the one God, his need to express himself, and the distinctiveness of each Person, Father, Son, and Spirit. We believe that we are made in the image and likeness of this God and therefore that we can be confident about what, in the proper circumstances, we can perceive about him from our experience. It is right for us to reflect upon the deepest relationships that we enjoy and which wondrously give our lives meaning, for through them we can come to understand, or perceive, or glimpse a little of his life. When we know and love most deeply, then are we drawn to share most intimately in that life also. We come to prayer.

(vii)

When Jesus prayed, he called his Father 'Abba'. That intimate familial usage was distinctively his. He taught and encouraged us to pray in the same way: 'Our Father, who art in heaven.' It is proper to do this, as we seek to live according to that fidelity in love which reveals that Jesus is our brother. But we know how feeble we are, and so we must seek help from the Spirit whom St Paul once cast engagingly as our interpreter: 'The Spirit helps us in our weakness; for we do not know how to pray as we ought, but the Spirit himself intercedes for us with sighs too deep for words. And he who searches the hearts of men knows what is the mind of the Spirit, because the Spirit intercedes for the saints according to the will of God' (Rom. 8: 26-7). The Church's liturgy is based on this pattern: the Father is worshipped through the Son and in the Holy Spirit. All our living, all our loving, all our praying must draw us to a deep sharing in the life of Father and Son and Holy Spirit so that we may reveal his image in our lives and in our world and at the end be united with him in glory.

> Batter my heart, three-personed God, for you
> As yet but knock, breathe, shine, and seek to mend;
> That I may rise and stand, o'erthrow me and bend
> Your force to break, blow, burn, and make me new.
> I, like an usurped town to another due,
> Labour to admit you, but O, to no end.
> Reason, your viceroy in me, me should defend,
> But is captived and proves weak or untrue.
> Yet dearly I love you and would be loved fain,
> But am betrothed unto your enemy.
> Divorce me, untie, or break that knot again,
> Take me to you, imprison me, for I,
> Except you enthrall me, never shall be free,
> Nor ever chaste except you ravish me.

(JOHN DONNE)

Further reading

Chapter 1

Butler, B. C., *An Approach to Christianity* (London, 1981).

Chapters 2–4

Brown, Raymond E., *Jesus God and Man* (London, 1968)
Vawter, Bruce, *This Man Jesus* (London, 1973)
Wansborough, Henry, *Risen From The Dead* (Slough, 1978).

Chapters 5–7

Brown, Raymond E., *The Critical Meaning of the Bible* (London, 1981)
——, *The Churches the Apostles Left Behind* (London, 1984)
Butler, B. C., *The Theology of Vatican II*, rev. edn. (London, 1981)
de Lubac, Henri, *Catholicism* (London, 1950)
Richards, Michael, *The Church of Christ* (Slough, 1982)
Sullivan, Francis, *Magisterium* (Dublin, 1983).

Chapters 8–10

Crichton, J. D., *The Once and Future Liturgy* (Dublin, 1977)
——, *Christian Celebration* (Leominster, 1981)
Hughes, Gerard J., *Moral Decisions* (London, 1980)
McCabe, Herbert, *The New Creation* (London, 1964)
Mahoney, John, *The Making of Moral Theology* (Oxford, 1986)
Quinn, James, *The Theology of the Eucharist* (Cork, 1973).

Chapters 11–13

Burtchaell, James T., *The Daily Dilemma of the Christian* (Chicago, 1973)
Coventry, John, *The Theology of Faith* (Cork, 1968)
Dominian, Jack, *Marriage, Faith and Love* (London, 1984)
Keane, Philip S., *Sexual Morality* (Dublin, 1980)
Simpson, Michael, *Death and Eternal Life* (Cork, 1971).

Chapter 14

Flanagan, Donal, *The Theology of Mary* (London, 1976)
Stacpoole, A. (ed.), *Mary's Place in Christian Dialogue* (Slough, 1982).

Chapter 15

Cantwell, Laurence, *The Theology of the Trinity* (Cork, 1968).

Index

Main discussions are indicated in italic.

OXFORD

MORE OXFORD PAPERBACKS

Details of a selection of other books follow. A complete list of Oxford Paperbacks, including The World's Classics, Twentieth-Century Classics, OPUS, Past Masters, Oxford Authors, Oxford Shakespeare, and Oxford Paperback Reference, is available in the UK from the General Publicity Department, Oxford University Press (JH), Walton Street, Oxford, OX2 6DP.

In the USA, complete lists are available from the Paperbacks Marketing Manager, Oxford University Press, 200 Madison Avenue, New York, NY 10016.

Oxford Paperbacks are available from all good bookshops. In case of difficulty, customers in the UK can order direct from Oxford University Press Bookshop, 116 High Street, Oxford, Freepost, OX1 4BR, enclosing full payment. Please add 10% of published price for postage and packing.

ROMAN CATHOLICISM IN ENGLAND

from the Elizabethan Settlement to the Second Vatican
Council

Edward Norman

'a brilliantly objective account . . . he has written about English
Catholicism in a manner for which English Catholics can be
grateful and of which he can be proud'

Lord Longford, *Contemporary Review*

'eruditely benign, fair, well-mannered and handling his theological, social and political researches with consummate ease.
Few scholars could take us from half-way through the Reformation to 1962 in fewer than 129 pages of text without
unbalancing history, but that is what the author has done.'

Sunday Telegraph

'full of insights . . . a model of clear and concise historical
writing' *Universe*

'a taut and sensitive history' *Church Times*

An OPUS book

CHRISTIANITY AND THE WORLD ORDER

E. R. Norman

This book, based on Dr Norman's 1978 Reith Lectures, considers a subject of great significance: the implications of the
contemporary politicization of Christianity. Ranging from the
political radicalism of Latin American Marxist Christians to
the problems encountered by Christianity in the Soviet Union,
Dr Norman identifies and presents a critical analysis of the
social and political ideas to which the modern Church is attach-
ing itself.

RELIGION AND THE PEOPLE OF WESTERN EUROPE 1789–1970

Hugh McLeod

In the years between the French Revolution and our own times there has been a widespread revolt against the various official churches that emerged triumphant from the turmoil of the Reformation and the Counter-Reformation. Religion became an integral part of the conflict between Right and Left, and economic change widened the gulf between the religious life of rich and poor.

Dr McLeod looks at the religious movements that flourished in these conditions and at the increasing difference between the religious life of the working class and that of the urban middle class in the growing towns and cities, and in the countryside. Finally, he considers how the religious patterns established in the period are gradually fading and changing.

'It is a wholly absorbing and valuable work, industriously researched, well written and with excellent notes and bibliography.' *Church Times*

An OPUS book

CHRISTIANITY IN THE WEST 1400–1700

John Bossy

This book aims to improve understanding of what happened to Western Europe at the time of the Reformation by renouncing the use of that term and its associated values. It also takes traditional or pre-Reformation Christianity seriously as an intelligible universe in its own right: slightly over half of the book is taken up by a systematic exposition of it. The second half explores the forces tending to undermine it, the characteristics of the regimes (Protestant and Catholic) which superseded it, and the consequences of its disintegration, in so far as it did disintegrate. Contrary to views now widely held, this book assumes that the population of the West consisted of Christians throughout the centuries in question (1400–1700), and that the social history of Europe and the history of Christianity were in this period substantially the same thing.

An OPUS book

ISLAM

H. A. R. Gibb

A knowledge of the tenets of Islam is crucial to our understanding of nations that have assumed an immense importance in the modern world. This book has become a standard work on a religion which is, after Christianity, the most widely diffused in the world. Beginning with Islam's origins as a practical religion in the Koran and the preaching of Muhammad, the author traces the growth of Islamic theology and the expansion of the Muslim social order, and the rise of the mystical Sufi movement. The concluding chapter analyses the problems that have confronted Islam in the twentieth century and its reactions to them.

An OPUS book

HINDUISM

R. C. Zaehner

Hinduism is both a way of life and a highly organized social and religious system, but in the modern world its essentials, perhaps more than those of any other major religion, are undergoing a process of redefinition. What then are the key concepts of Hinduism?

Professor Zaehner's book traces these through the four-thousand-year development of Hinduism, and is concerned to elucidate, as the author puts it, 'the changeless ground from which the proliferating jungle that seems to be Hinduism grows'.

'the best short introduction to Hinduism in existence . . . a really first-class guide to its ancient roots and to its subtle blend of the mythological and the metaphysical' *Guardian*

An OPUS Book

MUHAMMAD

Prophet and Statesman

W. Montgomery Watt

A short account of the life and achievements of one of the great figures of history, this volume also serves as an excellent introduction to one of the world's major religions.

Dr Watt tells of Muhammad's call to prophethood as a result of visions. He recounts the writing down of the Prophet's revelations in the Qur'ān (with an explanation of its passages); Muhammad's betrayal, expulsion from Mecca, and migration to Medina, and his rise to political power in Arabia. Throughout, Dr Watt makes clear the social and political background out of which Islam was born, especially the influence of Judaism and Christianity.

'This book . . . admirably fulfills its purpose. It is written in a clear and interesting style, and the reader can be assured that it is not only an interesting book to read, but is also based on sound scholarship.' *Journal of Semitic Studies*

WHAT IS THEOLOGY?

Maurice Wiles

'Professor Wiles's book makes a first-rate introduction to the subject. It has the honesty, the quiet persuasiveness and the penetration which we have come to associate with his work. Let those who undervalue theology read it and then ask themselves if theology is either a soft option or an irrelevant pastime.' *Times Literary Supplement*

'It is a lively, stimulating, and surprisingly thorough treatment, and one which certainly conveyed to me 'the worthwhileness and the excitement of the subject', which is indeed one of the author's aims.' *Theology*

AN INTRODUCTION TO THE PHILOSOPHY OF RELIGION

Brian Davies

Does rational inquiry show religious doctrines to be false, incoherent, or meaningless? Are there logical arguments for thinking that God exists or does not exist? And what, in any case, does 'God' mean? Does it make sense to postulate a good God, given the reality of evil? Does the idea of 'miracles' have any meaning? Can there be a rational basis for ethics which takes no account of God? Is the notion of human survival after death coherent?

This book is written for all who have been puzzled by these and similar problems, not just for students and professional philosophers. None of the questions is new, and Brian Davies examines critically the way they have been treated in the past by such philosophers as Anselm, Aquinas, Descartes, Leibniz, Hume, and Kant, as well as looking at the work of a number of modern thinkers.

An OPUS book

THE INTERPRETATION OF THE NEW TESTAMENT 1861–1961

Stephen Neill

'it is probably his attention to the personalities of the scholars involved in the debate which makes Bishop Stephen Neill's history of a century of New Testament interpretation of such absorbing interest. By any standard his grasp of the technical details of a highly complex subject is enviable, and the extent of his knowledge of the major works of New Testament criticism in five languages is astonishing . . . Altogether an excellent book.' *Times Literary Supplement*

THE CONCISE OXFORD DICTIONARY OF THE CHRISTIAN CHURCH

Edited by E. A. Livingstone

This is the abridged version of the second edition of *The Oxford Dictionary of the Christian Church*. It makes available for the general reader the vast majority of the entries in the parent volume. The range of the *Concise Dictionary* is considerable. It includes: the major Christian feasts and denominations, historical accounts of the lives of the saints, résumés of Patristic writings, and histories of heretical sects. It also outlines the opinions of major theologians and moral philosophers, and explores many related subjects.

Oxford Paperback Reference

NEWMAN

Owen Chadwick

The religious leader John Henry Newman started his long career as a devout Protestant; he later became the head of a new movement of Catholic ideas within the Church of England, and finally joined the Roman Catholic Church. He began a new epoch in the study of religious faith. Professor Chadwick examines the many aspects of Newman's thought and writings, especially his views about faith, knowledge and education.

'a fine introduction to the spirit of Newman' *Sunday Times*

...NAS

Anthony Kenny

Anthony Kenny writes about Thomas Aquinas as a philosopher, for readers who may not share Aquinas's theological interests and beliefs. He begins with an account of Aquinas's life and works, and assesses his importance for contemporary philosophy. The book is completed by more detailed examinations of Aquinas's metaphysical system and his philosophy of mind.

'It is hard see how such a book could be done better.' *London Review of Books*

JESUS

Humphrey Carpenter

Humphrey Carpenter writes about Jesus from the standpoint of a historian coming fresh to the subject without religious preconceptions. He examines the reliability of the Gospels, the originality of Jesus's teaching, and Jesus's view of himself. His highly readable book achieves a remarkable degree of objectivity about a subject which is deeply embedded in Western culture.

'Mr Carpenter has obviously made a thorough study of the latest New Testament scholarship: but he has also read the gospels with great care, pretending to himself that he was doing so without preconceptions, as a historian newly presented with the source-material . . . the most extraordinary achievement.' *Observer*